Misunderstanding science?

The public reconstruction of science and technology

Misunderstanding science?

The public reconstruction of science and technology

EDITED BY ALAN IRWIN
Reader in Sociology, Brunel University

AND

BRIAN WYNNE
Professor of Science Studies, Lancaster University

CAMBRIDGE
UNIVERSITY PRESS

Published by the Press Syndicate of the University of Cambridge
The Pitt Building, Trumpington Street, Cambridge CB2 1RP
40 West 20th Street, New York, NY 10011–4211, USA
10 Stamford Road, Oakleigh, Melbourne 3166, Australia

First published 1996

Printed in Great Britain by Redwood Books, Trowbridge, Wiltshire

A catalogue record for this book is available from the British Library

Library of Congress cataloguing in publication data

Misunderstanding science? The public reconstruction of science
and technology / edited by Alan Irwin and Brian Wynne.
p. cm.
Includes bibliographical references and index.
ISBN 0 521 43268 5 hardback
1. Science – Social aspects. 2. Science news. I. Irwin, Alan, 1955- . II. Wynne, Brian, 1947- .
Q175.5.M58 1996
306.4'5 – dc20   95-32980 CIP

ISBN 0 521 43268 5 hardback

Contents

Acknowledgements

The research in this volume was supported by the Economic and Social Research Council/Science Policy Support Group research programme in the 'public understanding of science'. The contributors wish to express their gratitude for the funding and assistance which was offered by this programme. Of course, this does not mean that either of these bodies should be held responsible for any errors or deficiencies in the accounts which follow. The editors would also like to thank Gillian Maude for her assistance.

Introduction

ALAN IRWIN AND BRIAN WYNNE

One of the most routine observations about modern life concerns the rapid pace of technical change and the consequences of this for every aspect of society. Of course, this is not just a phenomenon of the 1990s. The social impact of ceaselessly changing science and technology has been a classical theme of writers, social scientists and scientists since the Industrial Revolution. Generally, the tone has been deterministic, suggesting that science and technology have their own objective logic to which society must adapt as best it can.

However, the relationship between scientific expertise and the 'general public' is currently a matter of renewed attention and social concern. Although the dominant form of this renewed interest is shaped by anxieties about the 'social assimilation' of science and technology (i.e. by a concern that the public are insufficiently receptive to science and technology), we will argue that this conceals a more fundamental issue regarding the public identity and organisation of science within contemporary society.

This edited collection focuses on one important aspect of this wider theme; the contemporary issue of what has become known as the 'public understanding of science'. As the following chapters demonstrate, this has become something of a fulcrum for debates over the social negotiation of power and social order in relation to science and technology. In this Introduction, we will first set the scene for the detailed analyses which follow and then explain the particular approach to this debate which has been adopted here. As this book demonstrates, concern with 'public understanding' takes us into many areas of case-study and socio-technical inquiry – it is thus all the more important to establish from the outset the major interlinkages and connections.

The main themes of this book can best be introduced through some specific instances. Certainly, and as the chapters of this book will argue, the often-problematic relationship between 'expert knowledge' and the 'public' typically emerges in everyday life as part of particular issues.

The debate over civil nuclear power is often presented by the nuclear industry and government agencies as a division between nuclear 'experts' and an emotional public. Accordingly, public education is seen as the best way to win over support – if only people knew the facts then they would not worry unduly. However, this commitment to 'educating' the public is not just limited to the pro-nuclear lobby. Environmentalist groups also are keen to disseminate the 'real facts' about nuclear power. On each side, technical arguments are central to the debate. Meanwhile, the public are confronted with conflicting technical assessments of nuclear risk offered by groups who each claim a special understanding of the 'facts'. In each case also, these technical assessments represent an important part of the attempt to win over public opinion to a particular stance on the nuclear issue.

A similar analysis can be made of the 1990 debate over what became christened 'mad cow disease' (but known in scientific discussions as BSE – Bovine Spongiform Encephalopathy). Here, we had statements from the British Department of Health and also from distinguished figures such as Professor Sir Richard Southwood informing the public that the risks of BSE were tiny. As Sir Richard argued, 'we have more reason to be concerned about being struck by lightning than catching BSE from eating beef and other products from cattle.'[1]

Meanwhile, public concern was high – as indicated by the sudden drop in meat sales accompanied by a steep rise in media attention. Despite the official statements on BSE and the claims that scientific evidence suggested the risks to be small, two aspects of the public debate were very apparent. Firstly, that – as with the nuclear issue – scientific opinion was by no means unanimous (with Professor Richard Lacey, for example, taking a public stand against the 'official' position). Secondly, as the House of Commons Select Committee on Agriculture observed, 'Scientists do not automatically command public trust.'[2]

This was to become a very familiar message with regard to BSE. Accordingly, whilst there was much criticism of the general public for their 'emotive and irrational' response to the risks of BSE, we also begin to see that there may be some more complex social relationships at work (for example, concerning the basis of trust in scientific expertise). Nevertheless, what seems unavoidably true is that scientific argument was central to the 'mad cow' debate – with public groups and individuals being obliged to respond to the technical debate either by acceptance or rejection. Going further, we can discern that various forms of scientific evidence were used to defend public stances on BSE. We also see in a case like this that personal decisions must be taken in the face of conflicting technical claims and apparent uncertainties. Quite clearly, therefore, scientific arguments play an important role in structuring (or 'framing') the conduct of public debate. Equally, we can suggest that science is itself framed by unstated social commitments.

This role of science in 'framing' public debate, and the implicit social framing of science itself, will be a major theme of the coming chapters. We will argue that science in this way offers a framework which is unavoidably *social* as well as technical since in public domains scientific knowledge embodies implicit models or assumptions about the social world. In addition, as an intervention in public life, scientific knowledge involves rhetorical claims to the superiority of the scientific worldview but also it builds upon social processes of trust and credibility. Thus, whilst claiming to stand apart from the rest of society, science will reflect social interests and social assumptions.

Other examples of the 'public understanding of science in action' could readily be highlighted at this stage; debates over biotechnology and concerns over new reproductive technologies; advice on HIV/AIDS and 'safe sex'; discussions over information technology and its impact on jobs, skills, and the quality of life; pollution and hazard issues; global environmental change; medical problems, childbirth, and contraception; food safety and occupational health. In all these – and many more – areas, some kind of interchange exists between scientific assessments and public actions and responses. This very pervasiveness of 'expertise' makes the 'public understanding of science' an important area for discussion and analysis even though it means that the debate also becomes very broad-ranging, ill-defined, and at times slippery. Equally, we should note that various sociological accounts of our current social structure – often described as late- or new-modernity – stress the centrality and pervasiveness of technical expertise.[3]

Even more importantly, in all these areas social as well as technical judgements must be made – the 'facts' cannot stand apart from wider social, economic, and moral questions even if rhetorically they are often put forward as if this were the case. We can readily gather on the basis of the discussion so far that the relationship between science and the public may not be so straightforward as suggested in the conventional treatment which assumes a clear boundary between 'facts' and 'values'.

However, as has already been suggested, concern over the public understanding of science – either from the viewpoint of public groups or of scientists – is nothing new. Layton, for example, has shown how nineteenth-century concerns to advance public scientific literacy were imbued with an underlying anxiety to impart a particular worldview, one which would maintain social order and the legitimation of state institutions. Berg has analysed the Mechanics Institute movement in Britain (in the 1820s and 1830s) in similar terms.[4]

From another perspective, in the period around the Second World War the 'visible college' of left-wing scientists argued the need for a greater citizen awareness of science.[5] As J. B. S. Haldane put it in his 1939 book, *Science and Everyday Life*;

> The ordinary man must know something about various branches of science, for the same reason that the astronomer, even if his eyes are fixed on higher things must know about boots. The reason is that these matters affect his everyday life.[6]

Here we see one blunt statement of the public need to understand science – even if the 'higher things' pun also implies a notion of the inherent superiority of the scientific worldview over the shoemaker's craft (not an unusual nuance in scientific discussion of the 'ordinary person').

Writing immediately after the Second World War, the Association of Scientific Workers expressed similar sentiments in their programmatic *Science and the Nation*. Their argument for the scientific education of the public drew upon three of the most commonly stated justifications for an 'improved' public understanding:

- that a technically literate population is essential for future workforce requirements;
- that science is now an essential part of our cultural understanding;
- that greater public understanding of science is essential for a modern democracy . . .

> If responsibility . . . rests ultimately on the citizen – as in a democracy it must – then the citizen must be aware of and evaluate the technical as well as social aspects of the problem. Democracy needs a greater technical awareness, a rise in the standards of social and technical thinking.[7]

We notice here that in all this debate 'science' itself is constructed as unproblematic – its epistemic commitments, social purposes, institutional structures, intellectual boundaries and relationship with 'non-science'. This treatment of science has certainly been carried forward into the modern debate over 'public understanding'. A further issue which is central to this book – but which is generally concealed within this debate – concerns the meaning of 'understanding'. Most often, this is seen to equal faithful assimilation of the available scientific knowledges including their framing assumptions and commitments.

In order to pursue their goal of greater public understanding of science, the Association of Scientific Workers made a number of recommendations concerning the 'broadening' of education through further education classes and such media as exhibitions and museums, film, the press, and radio. They also made a plea for working scientists to become more involved in public activities.

Although there were sporadic outbursts in-between, the debate over 'public understanding of science' re-emerged particularly strongly in 1985 with the publication of a Royal Society report on the subject – suggesting both the durability of these issues and the perceived absence of substantial progress. It was also in the 1980s that the UK Royal Society and the American Association for the Advancement of Science formed their respective Committees for the Public Understanding of Science, thus institutionalising the subject.

The Royal Society considered the significance of this issue in terms which are highly reminiscent of the Association of Scientific Workers – except for the absence of socialist rhetoric. The Royal Society report instead presents itself as concerned with the general well-being of both science and society. However, the argument put forward by the Royal Society would surely have been endorsed by the Association of Scientific Workers:

> A basic thesis of this report is that the better public understanding of science can be a major element in promoting national prosperity, in raising the quality of public and private decision-making and in enriching the life of the individual . . . Improving the public understanding of science is an investment in the future, not a luxury to be indulged in if and when resources allow.[8]

This theme has been more recently endorsed by the UK Government in its 1993 White Paper on Science Policy.[9]

Since the Royal Society report served as a stimulus to the work in this volume – and since accordingly it will be referred to directly in a number of chapters – it is worth discussing it here in a little detail.

The Royal Society cites a number of specific areas where an 'improved understanding' would be of personal and national value. In many ways, this list represents an elaboration of the justifications given earlier by the Association of Scientific Workers. The need for a wider public understanding is justified in terms of:

- national prosperity (for example, a better-trained workforce),
- economic performance (for example, beneficial effect on innovation),
- public policy (informing public decisions),
- personal decisions (for example, over diet, tobacco or vaccination),
- everyday life (for example, understanding what goes on around us),
- risk and uncertainty (for example, concerning nuclear power or BSE),
- contemporary thought and culture (science as a rich area of human inquiry and discovery).

In each of these areas, improved technical understanding would enrich society and improve the quality of decision-making.

> Better overall understanding of science would, in our view, significantly improve the quality of public decision-making, *not* because the 'right' decisions would then be made, but because decisions made in the light of an adequate understanding of the issues are likely to be better than decisions made in the absence of such an understanding.[10]

This is, of course, a powerful argument which could be directly applied to the cases of BSE and nuclear power as previously discussed – a scientific understanding will illuminate the possibilities for action and allow a more considered response to everyday technical questions and problems. On this basis, the Royal Society advocated a series of changes within the education system, parliamentary bodies, the mass media, industry, and – especially – among scientists themselves in improving the current situation. Interestingly also, the Royal Society envisaged a role for *social* science research in this area – particularly in terms of gauging the present level of public understanding (or ignorance) of science, assessing the effects of improved understanding, and discovering *from where* individuals currently obtain technical advice and information. In an indirect fashion, the chapters in this collection represent a response to this demand for social scientific analysis. However, as we will now consider, the perspective adopted in this book diverges sharply from that of the Royal Society.

In particular, we can see certain assumptions embedded in the approaches to 'public understanding of science' considered so far – both as demonstrated by general accounts (such as that provided by the Royal Society) and also within contemporary controversies (for example, over BSE). First of all, there is an apparent assumption of 'public ignorance' in matters of science and technology – an assumption which has been bolstered by recent questionnaire surveys. According to these, the general public often lacks a basic understanding of scientific facts, theories and methodologies. Public controversy over technical issues is created by inadequate public understandings rather than the operation of science itself. This projection of a 'public ignorance' model also serves to problematise the general public rather than the operation of scientists and scientific institutions – just why aren't the public more responsive?

Secondly, there is an assumption that science is an important force for human improvement and that it offers a uniquely privileged view of the everyday world. Thus, the Nobel prize-winner Max Perutz approvingly quotes Nehru in his combatively entitled book, *Is Science Necessary?*:

> It is science alone that can solve the problems of hunger and poverty, of insanity and illiteracy, of superstition and deadening custom and tradition, of vast resources running to waste, of a rich country inhabited by starving people . . . Who indeed could afford to ignore science today? At every turn we have to seek its aid . . . The future belongs to science and those who make friends with science.[11]

Finally, science is portrayed in these accounts as if it were a value-free and neutral activity. Science *illuminates* and *assists* – it does not constrict or legitimate. Equally, the conditions under which scientific knowledges are constructed and validated are not challenged by the Royal Society. Science is unproblematically 'scientific' – it represents the *only* valid way of apprehending nature.

These points are, of course, important at a time when science is actually under criticism from a number of directions, and is generally suffering a marked lack of public support when it tries to rally a defence against outside attacks. Thus, whilst from the perspective presented so far science is the potential saviour of society, more critical voices have portrayed science as a major *cause* of ecological damage, military threats, constraints on personal liberty, and social disruption. In that sense also, it can be argued that the 'public understanding of science' represents an attempt by scientific institutions to regain their social standing and status in the face of public criticisms. As one of us has previously argued:

> the re-emergence of the public understanding of science issue in the mid-1980s can be seen as part of the scientific establishment's anxious response to a legitimation vacuum which threatened the well-being and social standing of science.[12]

This 'institutional neurosis' of science over public identification and legitimation is not by any means a new phenomenon. On the contrary, it appears to be a chronic condition throughout the history of science – albeit at a varying level of anxiety. The recent re-emergence of the issue thus calls for a deeper reflection than has so far taken place.

The chapters in this book seek to move the analysis of 'public understanding of science' away from the prevailing science-centred framework as sketched so far (and also from a simple oppositional or 'anti-science' stance). Instead of assuming that the problem is only or mainly with the public, we examine *both* the operation of scientific expertises/institutions and different 'publics' in relation to one another. This *relational* focus is especially important within the following accounts. In doing so, we interpret both 'science' and the 'general public' as diverse, shifting and often-diverging categories. We also adopt a critical-reflective stance on the current debates over public understanding of science in order to consider their motivation and under-lying concerns. The general argument in this book is that we need to rethink and reconceptualise the relationships between 'science' and the 'public' if we are to make progress at the level either of understanding or practical intervention.

This process of rethinking the public understanding of science should begin with our notion of *science*. Contrary to the kinds of division and contradiction which are found in such cases as civil nuclear power and BSE, the image of science which is generally presented within the 'public understanding of science' is of a unified, cleanly bounded, and clear body of knowledge and method. We also need to consider the nature not only of 'scientific institutions' (i.e. those bodies directly concerned with the funding, management, and implementation of science and technology), but also of the much larger category of institutions within society which draw upon or exploit science as a source of defence, legitimation or profit (for example, to return to the examples at the start of this chapter, the nuclear and food industries and the related government departments).

In this book, we will draw upon the last two decades of research within the sociology of scientific knowledge which has convincingly demonstrated the *socially negotiated* nature of science.[13] This resource is needed in order to examine the *varying* (i.e. heterogeneous) constructions and representations of science and to consider the relationship between these representations and the social institutions which employ scientific discourses and arguments. Thus, science will not be represented as a simple 'body of facts' or as a given 'method', but as a much more diffuse collection of institutions, areas of specialised knowledge and theoretical interpretations whose forms and boundaries are open to negotiation with other social institutions and forms of knowledge.

It will, therefore, be apparent in what follows that we are not counter-posing a homogeneous body of 'science' against a more diverse array of 'public understandings'. Instead, we portray both 'public' and 'scientific' knowledges as building upon wider commitments and assumptions. Implicit in our collection is that only a properly sociological approach to contemporary science can give us a real insight into the issues of 'public understanding'. Otherwise, we are doomed to a sterile and even counter-productive juxtaposition of 'science' against 'non-science' rather than an appreciation of the diversity and social interdependence of different forms of science, knowledge and expertise.

Of particular relevance within this book will be the manner in which scientific boundaries are established and maintained, i.e. the way in which 'science' is separated from 'non-science' or 'everyday knowledge'. By analysing contemporary science from this perspective, we can consider the different faces which science presents to the wider public. A key part of this is how 'constructions of society' (for example, tacit assumptions about users or audiences) are embedded within, and shape, scientific constructs. We can also examine how assumptions are made or decisions taken about *which* aspects of science to highlight to particular audiences. Going further, we can consider how different social groups recruit scientific arguments in order to support their case (a process which is quite evident in cases such as nuclear energy or mad cow disease). We can also consider how what counts as 'science' may be shaped by social relations and institutional structures so that the very constitution of science will reflect wider social interests.

Put simply, the research in this collection will move beyond a mere problematisation of *the public*. Instead, we will consider the operation of science in everyday situations – and, in particular, the different forms and representations of science which confront public groups. This point will be important in terms of the analysis which follows. It will also lead us to consider not just the 'public understanding of science' but also the scientific understanding of the public and the manner in which that latter understanding might be enhanced. We assert that this perspective is essential to the expressed goal of improving public uptake and 'understanding' of science, since without such a reflexive dimension scientific approaches to the 'public understanding' issue will only encourage public ambivalence or even alienation.

Rather than assuming from the beginning of discussion, therefore, that science unconditionally deserves privileged status, we need to consider just how relevant and important scientific understanding is within everyday life. To accept science as a key resource in public issues is radically different from accepting its automatic authority in *framing* what the issues are. Scientific approaches typically confuse these fundamentally different dimensions. This requires a problematising of what is actually meant by 'scientific understanding' in various contexts. We will also – especially in the concluding chapter – look again at the consequences of our new approach for the organisation of science.

This critical treatment of science will be matched in the following accounts by an awareness of the diversity of *public groups*. As we have already suggested, scientific statements about 'public understanding' tend to draw upon some notion of the typical citizen (i.e. Haldane's 'ordinary man'). Very little justification is given for this portrayal – instead, the public is portrayed as a homogeneous mass which needs to be rendered more receptive to the insights of science. The 'public' exist as an audience for science; they are an object rather than a subject. At this stage, we need to remember Raymond Williams' observation: 'there are in fact no masses, but only ways of seeing people as masses'.[14]

Rather than simply adopting this 'top–down' and dissemination-oriented model, much of the research in this book takes a very sensitive and careful look at the public*s* for science and considers their needs and interpretations. It also, crucially, is alive to the ways in which scientific knowledge frequently embodies tacit commitments about audiences or user-situations which may then serve as unnegotiated social prescriptions. Several of the chapters also examine non-scientific forms of knowledge and expertise (for example, those generated by direct and practical experience of scientific or technological systems) and their relationship to formalised understandings.

Viewed from this perspective, important issues of trust and credibility arise – why should we believe something just because it claims to be scientific? What kind of social relationship or identity is being tacitly proposed, or imposed, within scientific communications?

We also need to consider the ways in which personal understandings of the world (and previous experiences) fit together with scientific accounts – 'making sense' of new information is revealed to be a complex process which is likely to draw upon a series of sources. We see, too, that, specific publics are likely to be sceptical, critical or simply hostile to scientific statements – often because such statements seem to emerge from an idealised and inappropriate model of real world conditions. We may also find a resistance to the perceived social interests which are embedded in scientific statements – as when science is used for the legitimation of industrial practices (for example, the continued operation of hazardous industries). Taken together, we will replace the notion of 'public ignorance' with a much richer pattern of social relations and personal understandings.

In seeking to reconsider both 'science' and 'the public', we must also examine the kinds of *mediating institution* which currently convey scientific argumentation to the general public. What kinds of institution are these and how do they select the forms of technical appraisal for dissemination? These 'mediating institutions' take a number of forms – environmental groups, local industry, the mass media, the organ- isers of science exhibitions, government officials, doctors – but they offer important routes to the public's experience and understanding of science. What issues arise from the operation of these groups which deserve attention here? We must ask also whether these institutions stand apart from the conventional model of science–public relations or whether they can offer new, and perhaps more effective, patterns of cultural and knowledge relations.

Finally, each of the chapters has implications for *practical interventions* in this area. In the concluding chapter, we will address this theme directly – what are the policy implications of the new approach to 'public understanding' which has been developed in this collection?

Why then choose such an apparently cryptic title for this book? Of course, *Misun- derstanding science?* runs the risk of reinforcing the very notion that we have already tried so hard to dispel, i.e. that the problem is one of the public 'misunderstanding' science (what we can call the 'deficit' model as discussed, for example, in Chapters Two and Five). However, our title also suggests the opposite – that it is *science* which misunderstands both the public and itself. Given that most discussion in this area has stressed the former kind of misunderstanding, our collection will give par- ticular attention to the latter.

Our sub-title reinforces this emphasis by suggesting the active processes through which people 'reconstruct' technical information within everyday life. We will, there- fore, be particularly concerned with the different 'senses' which science presents and also with the way in which the 'cloth of meaning' (see Chapter Three) is woven and rewoven. For the contributors to this book, the 'reconstruction' of science within everyday life is an active and often-demanding process. As we will see, public groups often show great resourcefulness in carrying out this essential task – a resourcefulness which frequently makes scientific messages appear simplistic and one-dimensional.

Book structure

As has already been suggested, the chapters in this book are designed not to docu- ment nor explain the 'public misunderstanding of science', but to explore the specific contexts within which different kinds of technical judgement are reached by 'lay' publics. Put differently, we need to examine how different publics succeed or other- wise in 're-constructing' science as part of their own agendas. In tackling this, we will observe the limitations of scientific information in terms of everyday decisions and consider the role of social and political judgement in making practical use of science.

As we have argued so far, this citizen-oriented perspective offers considerably richer insights into the role of science within everyday life than the simplistic and technocratic assertion that 'more science must be better'. However, such a perspective requires considerable sensitivity to the knowledges and understandings possessed by citizens within specific situations and contexts. In order to develop this perspective, the various authors within this book have adopted a detailed, context-specific and local analysis of 'public understanding' rather than large-scale surveys or sweeping generalisations. However, a series of common analytical questions recur throughout these case-studies;

- what do people mean by 'science' and 'scientific expertise'?
- where do they turn for technical information and advice?
- what motivates them to do so?
- how do they select from, evaluate and use scientific information?
- how do they relate this expert advice to everyday experience and other forms of knowledge? What is involved in its integration at this level?

These questions will run through a series of detailed investigations – covering such areas as scientific researchers and policy-makers, medical patients and their families, communities around hazardous industrial sites, environmentalist groups, Cumbrian hill farmers coping with radioactive fall-out, museum staff, and visitors.

It should also be explained that these studies all began as part of one Economic and Social Research Council (ESRC) and Science Policy Support Group (SPSG) programme within the 'public understanding of science' (not least so that we can gratefully acknowledge the assistance and support provided especially by the SPSG). Thus, whilst the case-studies and research questions may vary from chapter to chapter, they do share a framework of analysis and understanding which has developed over a considerable period of time.

In stating this, we are aware that this range of study areas, and the very diffuseness of the 'public understanding of science' as an area of research and practice, may stretch the task of synthesis to the very limit. It would be quite wrong for us to pretend that all of these chapters fit neatly into a logical and coherent ordering. Whilst – as we have already argued – there is considerable thematic and analytical overlap in what follows, there is also a diversity of perspective and focus – not least because of the diversity of contributors to this book (see the 'Notes on contributors').

The area of analysis presented in this book contains a fresh approach and a new way of thinking about the sciences and their publics. It is only right, therefore, that there will be differences of emphasis and argument in what follows. We do not wish to conceal these or to consider their existence as an intellectual (or editorial) weakness – instead, *Misunderstanding science?* includes a range of contributors who, whilst agreeing on the broad framework of analysis, span a range of social scientific perspectives. We intend to offer in this collection both a fresh line of analysis and also a diversity of specific analyses. Furthermore, we feel it necessary to grasp something of the different social contexts within which the science–public relationship is encountered and negotiated.

Specific chapters within this collection focus on the nature of interactions between scientific representations and public responses. Others attempt to illustrate the constructed, contextual nature of scientific accounts. The final chapters look to the changing framework for science–public relations.

The first chapters present a strong set of 'contexts' for the public understanding of science. Brian Wynne presents an evocative example of one area of science–public interactions; the case of Cumbrian sheep-farmers in the wake of the Chernobyl disaster. Wynne analyses the relationship between the understandings and expertises held by the sheep-farmers and those of the official bodies who were attempting to control the sale and movement of radiation-contaminated sheep. Wynne raises questions of the construction and application of scientific knowledges in this specific context. He also discusses the robust views of local people when confronted with this kind of 'laboratory science'. The main points of this case-study are to elucidate the complex factors affecting the public credibility and uptake of scientific knowledge, and to highlight science as a culture involving its own pre-commitments and prescriptions beneath specific claims and 'facts'. Wynne argues that this is a general condition and that public responses originate at this level as well as that of specific claims.

The following chapter develops some similar arguments in a very different social context; the case of local communities living with the routine risk and pollution from the chemical industry. Even in this less coherent and less tightly defined social setting, Irwin, Dale, and Smith discover a critical public response to technical advice and information. Once again, and very importantly for this collection, science 'disappears' into the everyday life of this community – on the face of it, the official communications and public reactions have little or nothing to do with science, nor indeed does science figure as a part of everyday discussion about local hazards. However, as the authors argue, implicit models of technical expertise are employed to legitimate industrial assessments of risk. Such expertise is then met with a questioning and often sceptical local audience.

This 'disappearance' of science within specific contexts of action and understanding is a recurrent theme of *Misunderstanding science?* From the perspective of influential scientific institutions such as the Royal Society, science should be centre-stage within important areas of everyday life such as issues of risk and pollution. Certainly, science is drawn upon by industrial and government organisations as a source of rhetorical and ideological support for particular social practices – for example, in the case of Chapter Two, with regard to the operation of hazardous industries close to areas of housing. However, from the viewpoint of citizen groups, science may be much less significant than matters of trust, powerlessness and contextually generated expertises. The 'disappearance' of science does not mean that it serves an unimportant role in such situations – it is more that 'science' as a category blurs into other areas of social practice and contestation.

Within this context, the credibility and perceived 'knowledgeability' of different local information sources become an important point of community discussion and response. As with the sheep-farmers case, we see the publics of science to be active rather than passive in their response to official information. The publics are engaged in defining their relationships with science – demarcating its meanings, boundaries, and utility within their particular social situation. In both cases also, we see the practical difficulties and uncertainties experienced by science when applied to specific social contexts.

The third and fourth chapters of *Misunderstanding science?* focus on the public response to one very immediate and personal area of encounter with expertise – medicine and medical science. Lambert and Rose consider the ways in which patients with a particular genetic metabolic disorder 'make sense' of the medical science which they encounter. As the authors suggest, the 'scientific' nature of this interaction is all the greater since the disorder generally does not produce obvious external signs of 'disease'. Instead, those with this condition must somehow 'weave the cloth of meaning' from various expert accounts. The study accordingly focuses on the integration of science into everyday life and, in common with many other studies in this collection, examines the contested, provisional, and situated nature of scientific accounts. Despite this, where a public group has a strong motivation to acquire information, we see just how skilled and resourceful it can be in sifting through these accounts in order to 'make sense of science'. Lambert and Rose also demonstrate that in so doing public groups effectively renegotiate the boundary between science and the public as they come to take responsibility for aspects of knowledge previously taken for granted as the domain of science or medicine.

In Chapter Four of this collection, Frances Price offers a second medically related case-study. Her chapter is especially concerned with imaging technology within reproductive medicine and with the ways in which the enhanced 'visibility' of embryos and fetuses comes to be framed within antenatal and infertility clinics. The production and communication of these new sophisticated images can be portrayed as straightforwardly beneficial to patients – who can now 'see' into their own womb. However, as the chapter discusses in some depth, the production of such images for viewing by, and discussion with, patients also represents a shaping of knowledge and of the communication of knowledge which can raise difficult questions of authority, expertise, and control.

Chapters Five and Six are both concerned with ignorance – but show 'ignorance' to be actively rather than passively constructed. As Michael discusses, the conventional approach defines ignorance as an empty space, as a knowledge vacuum. However, Michael notes to the contrary that specific varieties of ignorance were demonstrated in his interviews with lay people using expressions of ignorance to structure their relationship to science. The author explores the *discourses of ignorance* that people mobilise when commenting on science. Drawing upon three areas of fieldwork, Michael notes that these discourses can take several identifiable forms – including ignorance as a deliberate choice – and that these will represent a reflection of the power relation between people and science. Michael's analysis suggests the importance of social identity as the fundamental basis of public responses to 'science' (which is itself being projected according to the processes of social identity maintenance of scientific and technical institutions).

Chapter Six offers a study of the reception of technical expertise within the culturally coherent community of the Isle of Man. In representing this on the basis of ethnographic investigation, McKechnie provides an account of the same fundamental processes described by Michael but in a very different social and cultural setting. Whereas the lay-publics investigated by Michael were relatively amorphous, not united by any collective identity, McKechnie studied the reconfiguration of external scientific expertise according to the relatively well-structured Manx cultural identity. The chapter argues, like that of Wynne earlier, that the processes of interaction with science were inextricably interwoven with wider concerns to defend this cultural identity against 'outsiders' who were seen to be using science as a political weapon. Once more, we enter the territory of 'reconstructing science' – with informal cultural frameworks acting as a medium of conditional and qualified assessment. Other basic processes involved at this level include; evaluation according to cultural and social criteria; social positioning and bounding according to prevailing or desired social relationships; distancing, reformulation, and adoption of information in more local idioms and languages. Although from a 'science-centred' perspective, science may appear to have been ignored, rejected or distorted, this chapter yet again reveals the more complex, sophisticated, and less-negative nature of public appreciations of science.

In Chapter Seven, we move to the construction of one form of scientific dissemination – the preparation of a permanent gallery on food within the Science Museum. The emphasis here is on the framing of science and on the processes of selection and audience-establishment involved. Through an ethnographic investigation (a methodology which is common to most of the chapters in this book), MacDonald records the social processes involved in the contextualisation of science and the manner in which the competing objectives of the exhibition are reconciled. Thus, MacDonald analyses the gallery's particular construction of contentious issues such as food safety and the manner in which the 'official' nature of the museum serves to inhibit controversy and dispute between experts. Overall, MacDonald argues powerfully that science communication involves considerably more than simply 'transporting information' from one medium to another – instead, it involves a definition and selection of just which 'facts' are to be presented to the public, of how 'problems' are identified and a negotiation over what kind of entity 'science' should appear to be.

Chapters Eight and Nine of this collection move discussion on to the changing context of science–public relations. We look at two major discussions here: the growth in importance of social action and organisation outside of conventional institutions, and the transformation of scientific cultures themselves towards more commercial forms.

First of all, Yearley addresses the important relationship between science and environmental organisations. Yearley suggests that the study of knowledge and beliefs about the environment offers a pressing example of the handling of science in the public realm. The author notes the ambivalent relationship between science in this context and practical policy-making and also the difficulties faced by green groups in making sense of science. From an analysis of the practical experiences of environmental organisations, it would appear that there is no straightforward way of harnessing scientific expertise to environmental interests. Equally, a growing awareness of the problems encountered by science in this area is leading to increasing challenge against science as a 'speaker of truth against power'. Such an awareness would have major consequences for the perceived importance of the public understanding of science.

Chapter Nine follows this general analysis of the changing relationship between public groups and science with a study of the changing nature of *science* itself. Rothman, Glasner, and Adams explore the differing models of 'science' held by those working within scientific institutions – including scientists, R and D managers, research council officials and others dealing in strategic science. They argue that there is no clear consensus amongst the practitioners of science concerning how 'science' or 'scientific knowledge' should be defined in any context. The authors thus suggest the diverse character of science as it is encountered by various publics. Science is currently undergoing a process of institutional change. For all these reasons, the conception of what is involved in the public understanding of science will of necessity become more heterogeneous and pluralistic.

In the Conclusion to this collection, we try not to summarise the various key points made by contributors but instead to address certain underlying and highly consequential issues – including the wider relevance of these largely qualitative and case-study oriented accounts. Is it possible to reach satisfactory conclusions for the 'public understanding of science' from a series of such specific contexts? This will also involve a methodologically oriented discussion of the relationship between such qualitative-interpretive research and the large-scale surveys which have investigated public understanding and attitudes towards science from a rather different perspective.

We will also attempt to locate the content of this book within the changing structure of contemporary society and, especially, debates over the nature of modernity and the cultural dimensions of political change. Finally, and very importantly, we will consider the practical implications of this line of analysis. What does our account tell us about the possibilities for enhancing future social relationships between citizens and science? These implications will relate both to public groups themselves and to the contemporary institutional structure of science. We aspire here to addressing the problems of legitimation which underlie the rhetorical expression of the 'public understanding of science' problem.

As we hope to demonstrate in the following chapters, the 'public understanding of science' represents an issue which is both practically pressing and intellectually challenging. We do not claim to have a tidy synthesis which can easily replace current 'science-centred' accounts and assumptions. Instead, our argument is that both practical policy-making and academic analysis needs to embrace the uncertainty, complexity and heterogeneity of current science–public interactions. As we will suggest, it is only on that basis that real social and scientific progress can be made.

Finally in this Introduction, we need to address questions of the audience for *Misunderstanding science*? If the arguments of this book are indeed as significant as we, the editors, consider them to be, then our readership needs also to be broad – both in terms of national and international coverage, and in terms of the backgrounds of individual readers. Whilst the specific studies presented in this book largely relate to the British context, we quite clearly see them as having a wider relevance. This point will be dealt with especially in the Conclusions where we tackle directly matters of the 'local' and the 'cosmopolitan'. Meanwhile, each of the chapters attempts to draw broad implications from each of the particular areas of investigation in a manner which explicitly connects individual contexts to wider social themes, issues, and concerns. Thus, we can hypothesise that, whilst in other national settings the local contexts may indeed change, the same underlying questions will occur. Of course, this book is just a starting-point to that vital process of international inquiry.

As for the readers themselves, we hope that this book will fit within social science-based discussions around the sociology of modernity and of scientific knowledge and interdisciplinary treatments of science and technology studies. However, we intend also that this book will be accessible to that larger audience of scientists and technologists (whether working in a research capacity or within what we can loosely term 'scientific institutions') and of public groups who are in different ways grappling with the issues of 'misunderstanding science?'.

We recognise that the language of this book will at times appear a little difficult and 'technical' – it is important that social scientists should be able to use specialised concepts and the language of theory (although, as editors, we have done our best to keep this to a minimum). It is our contention that the issues discussed here demand a high order of analysis – if that analysis occasionally necessitates the use of difficult concepts then that is reasonable. At no point, however, should that conceptual complexity interfere with the central aim of this book; to explore, analyse, and (as a consequence) reinterpret the problematic relationship between the sciences and their publics in modern society.

NOTES

1. Southwood, R., quoted in Food Safety Advisory Centre leaflet, *The Facts about BSE*. The Food Safety Advisory Centre is sponsored by Asda, Gateway, Morrisons, Safeway, Sainsburys, and Tesco.
2. Quoted in *The Guardian* 13 July 1990.
3. See, for example: Beck, U., *Risk Society; towards a new modernity* (London, Newbury Park, New Delhi: Sage, 1992); Giddens, A., *Modernity and Self-Identity; self and society in the late modern age* (Cambridge: Polity Press, 1991).
4. Berg, M., *The Machinery Question and the Making of Political Economy 1815–1848* (Cambridge University Press, 1980).
5. See, Werskey, G., *The Visible College – a collective biography of British scientists and socialists of the 1930s* (London: Free Association Books, 1988).
6. Haldane, J. B. S. *Science and Everyday Life* (Harmondsworth: Pelican Books, 1939) p. 7.
7. Members of the Association of Scientific Workers, *Science and the Nation* (Harmondsworth: Penguin, 1947) p. 246.
8. Royal Society, *The Public Understanding of Science* (London: Royal Society, 1985) p. 9.
9. UK Government, *Realising Our Potential; a strategy for science, engineering and technology* Cm 2250. (London: HMSO, 1993).
10. Ibid.
11. Nehru quoted in Perutz, M., *Is Science Necessary? Essays on science and scientists* (Oxford and New York: Oxford University Press, 1991) p. vii.
12. Wynne, B., 'Public understanding of science research; new horizons or hall of mirrors?', *Public Understanding of Science* 1, 1 (1992) p. 38.
13. See, for example: Latour, B., *Science in Action* (Milton Keynes: Open University Press, 1987); Mulkay, M., *Sociology of Science; a sociological pilgrimage* (Milton Keynes: Open University Press, 1991); Woolgar, S., *Science: the very idea* (London: Tavistock, 1988).
14. Williams, R., *Resources of Hope*. Edited by R. Gable. (London and New York: Verso, 1989) p. 11.

1 Misunderstood misunderstandings: social identities and public uptake of science

BRIAN WYNNE

This chapter takes as its focus one very specific example of public interaction with science – the case of the hill sheep-farmers of the Lake District of northern England. They experienced radioactive fall-out from the 1986 Chernobyl accident which contaminated their sheep flocks and upland pastures. In an area dominated by a traditional and demanding hill-farming economy, they were restricted from selling their sheep freely (almost 100 farms are still under restriction). They also received intensive expert advice about the environmental hazards from the radiocaesium deposits, and the relationship of these to other such deposits from the nearby Windscale-Sellafield nuclear facilities and 1950s weapons testing fall-out. Fieldwork comprising mainly in-depth interviews with affected farmers and others provided data for analysis of the factors influencing the reception of scientific expertise by this sub-population.

In analysing the farmers' understanding of the science, it was immediately apparent that it would have been meaningless and utterly misleading to treat their response to its cognitive content – for example, the claim that radiocaesium from Chernobyl was clearly distinguishable from Sellafield emissions of the same radio-isotopes – as if separate from its social and institutional form. Examining how the scientific institutions framed the issue and the knowledge they articulated as science, identified certain commitments which were institutionalised and taken for granted, thus not deliberately introduced. They constituted the very culture of science as institutionalised and practised as public knowledge. These assumptions *shaped* the scientific knowledge, they were not *extra* to it; and they were built in as social prescriptions in the way the science was institutionalised and deployed. These were elements of cultural prescription posing, albeit innocently, as objective knowledge. They included the assumptions:

- that the natural (and achievable) purpose of knowledge was control and prediction;
- that standardisation of environmental measurements and concepts over given areas and social units was natural even though it imposed standardisation on the social units too;

- that uncertainties in scientific knowledge could be contained within the private discourse of the scientists and would be misunderstood if disclosed in public;
- that local lay knowledge was effectively worthless; and
- that scientific methods of research could fully simulate realistic-farming conditions as practised, transmitted, and valued in hill-farming culture.

These assumptions were embedded in scientific interventions and approaches. They were by no means neutral in relation not only to local culture, but to locally validated knowledges. With such expert interventions, people experienced their identities to be threatened. This was not remotely characterisable as a cognitive challenge, which is how conventional approaches would treat it (Royal Society 1985; Durant et al. 1989).

One central issue within this chapter will be that of trust. Certainly, it is now accepted that trust and credibility are major contextual factors influencing the uptake and understanding of scientific messages, and the public perception of risks (Wynne 1980, 1992; Slovic 1992). In particular, we need to consider the trust which public groups are prepared to grant to various institutions and actors – Chapter Two, for example, discusses this with particular reference to the relationship between the local chemical industry and residents over hazard issues.

In this chapter, these cultural questions of trust and credibility will be examined – not as intrinsic or inevitable characteristics of knowledge or institutions – but as embedded within changing social relationships. Thus, though they pervade all processes of 'understanding', trust and credibility are contingent variables which depend upon evolving relationships and identities. As such, lay judgements of trust are not set in concrete or even necessarily apply in all circumstances. Rather, they are conditional, and open to continual renegotiation. Indeed, as the following case-study suggests, even in the supposedly traditionalist culture of hill farming these judgements may be subject to ambivalence and contradiction – reflecting the diverse social networks and multi-faceted identities inhabited by this one social group. In recognising the cultural role of trust relationships therefore, we should avoid the reification of trust, and retain its problematic conceptual character.

Throughout this case-study, we will consider the interplay between social and cultural identities (especially those of the sheep-farmers) as they see themselves threatened by the form of scientific interventions. All too often, such confrontations are portrayed as a clash between open-minded modernity and closed-minded tradition. By contrast, in what follows, we will witness not only the grounded and reflexive cognitive content of lay knowledges, but also the culturally-loaded structures of scientific knowledge as deployed in public domains (Michael 1992; Wynne 1993).

It is not, therefore, that scientific knowledge merely *omits* social dimensions that ordinary people incorporate in their evaluations and assessments. It is that scientific knowledge tacitly imports and imposes particular and problematic versions of social relationships and identities. This seems a major factor in the sometimes negative public response to technical pronouncements, especially ones which, in their lack of institutional self-awareness or reflexivity, impose these social prescriptions without negotiation. This will be a recurrent theme in the kinds of social circumstance analysed in this book.

Thus, three key points of wider significance are apparent:

- the fundamental interaction between scientific expertise and lay-publics is *cultural*, in that scientific knowledge embodies social and cultural prescriptions in its very structure;
- the problems of public uptake of science therefore lie in the institutional forms of science and of its incorporation into policy and administration;
- 'local' case studies of this sort should be seen as an expression of deeper problems of modernity as embodied in dominant institutional cultures. They are not just a defence of 'traditional communities' against an anonymous modernising 'centre', but a more fundamental challenge to the very idea of a universalising 'centre' in the first place.

Sheep farmers, scientists, and radiation hazards: the background

The hill sheep-farmers near the Sellafield (formerly Windscale) nuclear fuels reprocessing complex in the Lake District of Cumbria, northern England, have more than a personal health interest in radiation risk information. Their economic viability depends totally upon rearing a large crop of lambs each spring, and selling them in the late summer and autumn, before they run out of their meagre valley grazing due to the temporary overpopulation of lambs. The UK lamb industry exports heavily to continental Europe. Any public perception of radioactive blight on its product would destroy the industry, especially the hill sheep-farming sector which is a key early part of the breeding cycle, but which is economically more fragile and offers the farmer few or no alternatives compared to lowland sheep-farming.

The upland hill-farming region in the Lake District is one of the few locations of relative solidarity and distinctive traditional cultural identity left in industrial Britain. Although (as shown later) this should not be overstated, these communities share an unusually demanding livelihood as a way of life; they occupy a distinct and sought-after geographical location, and have common historical traditions, linguistic dialects and recreational pursuits. They also share particular 'external' socio-economic threats such as subordination to tourism, landlords, and authorities who appear to be more and more concerned with meeting environmental and urban recreational demands on the country than with sheep-farming. All of this was an important context of the post-Chernobyl crisis.

In May 1986, following the Chernobyl accident, upland areas of Britain suffered heavy but highly variable deposits of radioactive caesium isotopes, which were rained out by localised thunderstorms. The effects of this radioactive fall-out were immediately dismissed by scientists and political leaders as negligible, but after six weeks, on 20th June 1986, a ban was suddenly placed on the movement and slaughter of sheep from some of these areas, including Cumbria.

Although this shock was mitigated somewhat by the confident scientific reassurances that the elevated levels of caesium in sheep, and hence the ban, would only last about three weeks, at the end of this period the restrictions were instead imposed indefinitely. The confident dismissal of any effects only two months earlier had changed to the possibility of wholesale slaughter and complete ruin of hill sheep-farms at the hands of a faraway stricken nuclear plant. At the time over four thousand British farms were restricted. The initially wide restricted area in Cumbria (which included about 500 farms) was whittled down within three months to a central crescent covering 150 farms (see Figure 1.1). These farms remained restricted, contrary to all the scientific assertions of the time. A later review indicated that they could remain so for years, overturning completely the scientific basis upon which the previous policy commitments were made (Howard and Beresford, 1989).

Very close to this recalcitrant central 'crescent' of longer-term radioactive contamination, almost suggesting itself as its focal point, is the Sellafield-Windscale nuclear complex. The stories of Sellafield-Windscale and Chernobyl are intertwined in ways which I now unravel.

Sellafield-Windscale is a huge complex of fuel storage ponds, chemical reprocessing plants, nuclear reactors, defunct military piles, plutonium processing and storage facilities, and waste processing and storage silos. It has developed from its original role in the early 1950s of producing purely weapons-grade plutonium into a combined military and commercial reprocessing facility which stores and reprocesses thousands of tonnes of UK and foreign spent fuel. It is by far the biggest employer in the area, with a regular workforce of some five thousand swollen until recently by a construction workforce of nearly the same size. It dominates the whole area not only economically, but also socially and culturally.

Sellafield has been the centre of successive controversies, accidents, and events relating to its environmental discharges and workforce radiation doses, with increasingly powerful criticisms not only of allegedly inadequate management and regulation, but also of poor scientific understanding of its environmental effects, and of the economic irrationality of the recycling option in nuclear fuel cycle policy. In the early 1980s the plant was alleged to be the cause of excess childhood leukaemia clusters; these excesses were confirmed in 1984 by an official inquiry chaired by Sir Douglas Black, which nevertheless expressed agnosticism as to the cause (Macgill 1987; McSorley 1990). This controversy continues, with every scientific report

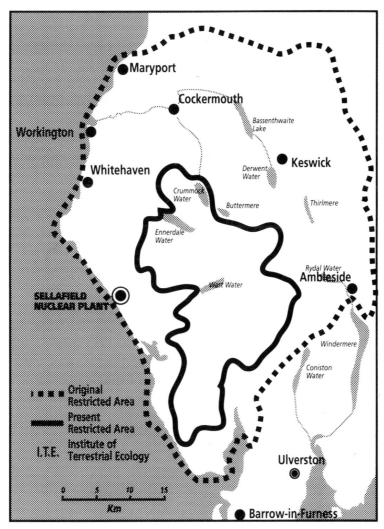

1.1 Map showing the restricted areas of Cumbria, from June 1986 (the original area), and from September 1986 (the present one). (Source: Drawn from UK Ministry of Agriculture, Fisheries and Food (MAFF))

exhaustively covered in the local and national media (Gardner et al. 1990). The plant operators were later shown to have misled the Black inquiry, inadvertently or not, over earlier levels of environmental discharge of radioactivity. In 1984 the operators were accused by the environmental group Greenpeace of contaminating local beaches above legal discharge levels, and were subsequently prosecuted; and in 1986 they were threatened with closure after another incident and an ensuing formal safety audit by the Health and Safety Executive. Despite huge investments in public relations, they have suffered a generally poor public image for openness and honesty over the years.

Before most of these controversies developed, in 1957 the Sellafield-Windscale site suffered the world's worst nuclear reactor accident before Chernobyl, when a nuclear pile caught fire and burned for some days before being quenched (Arnold 1992). It emitted a plume of radioactive isotopes, mainly iodine and caesium, over much the same area of the Lake District of Western Cumbria as that affected by the Chernobyl fallout. This fire and its environmental effects were surrounded by a great deal of secrecy. Although farmers in the vicinity were forced to pour away contaminated milk for several weeks afterwards, at the time they reacted without any overt hostility or criticism of the industry. Even in 1977 when they had the opportunity during a public inquiry to join with an emergent coalition of environmental groups against a major expansion at Sellafield, the local farming population largely kept out of the argument (Wynne 1982). Significantly, however, it was only after 1987 that the fuller extent of the Windscale fire cover-up emerged into the public domain. In 1990 it was revealed in a television programme that the ill-fated pile had in fact been allowed to operate with faults, which meant that highly irradiated spent fuel had been lying in the air streams emitted up the stack. Thus it was exposed that the fire had been a blessing in disguise for the authorities, since any discoveries of local environmental contamination could be attributed to the one-off fire itself rather than to longer-standing management practices which had allowed routine uncontrolled radioactive emissions to occur for some time before. The parallels with the Chernobyl issue nearly thirty years later are remarkable, as explained below.

The farming population in the Cumbrian hills is relatively stable, most farmers having lived through these controversies and events as near neighbours. Indeed many of them have relations, neighbours, and casual employees who also work at the Sellafield-Windscale site. Not only is it close physically, it is also never far away from contemplation. Far from giving Sellafield-Windscale some welcome relief, the Chernobyl emergency ironically brought it even more to critical public attention.

Scientific knowledge and social identities

At first, scientific advice was that there would be no effects at all from the Chernobyl radiocaesium fall-out. After six weeks these confident public reassurances were dramatically overturned when on 20 June 1986 the Minister for Agriculture announced the complete ban on sheep sales and movements in several affected areas, including Cumbria (see Figure 1.1). However, the shock waves from this reversal were contained by the strong reassurances accompanying the ban that it would be for three weeks only, by which time radioactivity levels in lamb would, it was confidently expected, have reduced beneath the level at which intervention was required. This short ban could be accommodated because very few if any hill lambs would be ready to sell before late August anyway.

Yet after the three-week period, instead of lifting the ban the government announced an indefinite extension, albeit for a smaller area. This represented an altogether more serious situation in which the hill farmers faced ruin, because not only lamb crops, but also breeding flocks faced starvation and wholesale slaughter due to inadequate grazing. The government introduced a scheme to remove this threat: it allowed farmers to sell lambs contaminated above the limit if they were marked, in which case the lambs could be moved to other areas but not slaughtered until their contamination had reduced. This blight factor collapsed the market price for marked sheep, and many lowland farmers bought them and then made handsome profits when they sold them after the sheep had decontaminated on their farms.

The hill farmers were left in a quandary. If they sold, they had to run the gauntlet of the threatening bureaucratic system which had been established to manage the restrictions, which consisted of prior notification, tests and controls, and paperwork, and offered only a possible and partial future compensation for catastrophically low prices. If they held on to their sheep they risked ruin from starvation, disease, and knock-on effects, or from the costs for buying in extra feed. Yet, even after the initial contradiction of their scientific beliefs, the scientists advised farmers to hang on because, as they persisted in believing, the contamination would fall soon – it was just taking a bit longer than expected. When farmers did follow expert advice and waited, they found the restrictions continuing; once again the advice was badly over-optimistic, and had led them into a blind alley in which many costly complications to farm-management cycles had been introduced, and compensation was cut off because they had not sold within the prescribed period. In the circumstances it was not surprising that our interviews (see later) found many farmers bitterly accusing the scientists of being involved in a conspiracy with a government which they saw as bent on undermining hill farming anyway.

Through the troubled and confused summer of 1986, in spite of mounting evidence and their own public embarrassment, the scientists persisted in their belief that the initially high caesium levels would fall soon. Only later did it emerge that these predictions were based upon a false scientific model of the behaviour of caesium in the upland environment. The prevailing scientific model was drawn from empirical observation of alkaline clay soils, in which caesium is chemically adsorbed and immobilised and so is unable to pass into vegetation. But alkaline clay soils are not found in upland areas, which have acid peaty soil. This type of soil had been examined, but only for physical parameters such as depth-leaching and erosion, and not for chemical mobility (Wynne 1992).

Thus the scientists unwittingly transferred knowledge of the clay soils to acid peaty soil, in which caesium remains chemically mobile and available to be taken up by plant roots. Whereas their model had caesium being deposited, washed into the soil and then locked up by chemical adsorption, thus only contaminating the lambs on a one-pass basis, in fact the caesium was recycling back from the soil into vegetation, and thence back into the lambs. This mistake only became apparent over the next two years, as contamination levels remained stubbornly high and the reasons were urgently sought. What was not lost on the farmers, however, was that the scientists had made unqualified reassuring assertions then been proven mistaken, and had not even admitted making a serious mistake. Their exaggerated sense of certainty and arrogance was a major factor in undermining the scientists' credibility with the farmers on other issues such as the source of the contamination. In any case the typical scientific idiom of certainty and control was culturally discordant with the farmers, whose whole cultural ethos routinely accepted uncertainty and the need for flexible adaptation rather than prediction and control.

The structure of the scientific knowledge in play also embodied and, in effect, prescribed a particular social construct of the farmers (Wynne 1992). To summarise this analysis, the degree of certainty expressed in scientific statements denied the ability of the farmers to cope with ignorance and lack of control: and the degree of standardisation and aggregation of the scientific knowledge, for example the spatial units of variation of variables such as caesium contamination, denied the differences between farms, even in a single valley (and even within the same farm). At the same time the scientists ignored farmers' own knowledge of their local environments, hill-sheep characteristics, and hill-farming management realities such as the impossibility of grazing flocks all on cleaner valley grass, and the difficulties of gathering sheep from open fells for tests.

As a result the farmers felt their social identity as specialists within their own sphere, with its adaptive, informal cultural idiom, to be denigrated and threatened by this treatment. This was a reflection of the culture and institutional form of science, not only of what specifically it claimed to know.

A graphic example of the scientists' denial of the specialist knowledge of the farmers was the scientists' decision to perform experiments on the value of the mineral bentonite to chemically adsorb caesium in the soil and vegetation, thus helping reduce recontamination of grazing sheep (Beresford et al., 1989). The bentonite was spread at different concentrations on the ground in different plots; the sheep from each plot were then tested at intervals, and compared with controls on zero-bentonite land. However, in order to do this the sheep were fenced in on contained fell-side plots. The farmers pointed out that the sheep were used to roaming over open tracts of fell land, without even fences between farms, and that if they were fenced in they would waste (lose condition), thus ruining the experiment. Their criticisms were ignored, but were vindicated later when the experiments were quietly abandoned for the reasons that the farmers had identified. The farmers had expressed valued and useful specialist knowledge for the conduct and development of science, but this was ignored. Similar experiences occurred over other aspects of hill farming and the scientific knowledge relating to the management of the crisis.

In respect of both the 'conspiracy theory' and the 'arrogance theory' of science, the Cumbrian sheep-farmers felt that their social identity as a specialist community with distinct traditions, skills, and social relations was under fundamental threat. These two models of science, which reinforced each other in the experienced threat to social identity, are mutually contradictory if taken literally. The former implies omniscience ('they knew all along that the high levels would last much longer than they admitted'); the latter implies unadmitted ignorance in science. This apparent anomaly exposes the futility of expecting consistent formal 'beliefs' about science in research on public understanding; coherence lies at a deeper level, of the defence and negotiation of social identities. We examine this dimension next.

Public belief and private dissent

Before the Chernobyl plume deposited its radioactive burden on the fells of Cumbria, there had been little or no controversy about radioactive caesium and related contamination of sheep on the high fells. Amongst several other issues concerning Sellafield's environmental discharges, contamination of pastures and grazing animals along the lowland coastal plain near the plant had been found and debated, for example in monitoring by Friends of the Earth, the Sellafield operators British Nuclear Fuels, and the Ministry of Agriculture, Fisheries and Food (Friends of the Earth 1987; BNFL 1987; MAFF, 1987; HC Agriculture Select Committee 1988). But little or no scientific monitoring or public attention had been paid to the high fells and their sheep; and no allegations of contamination of the fells and their sheep had been made. When the Chernobyl restrictions were announced however, and then almost immediately extended indefinitely, questions were very soon circulating locally about the real source of the contamination. The fact that a crescent of high contamination almost centring on Sellafield persisted (Figure 1.1) against scientific reassurances that levels would decrease within a few weeks, was *prima-facie* evidence of a hitherto hidden Sellafield dimension. The first national maps of caesium contamination measured after Chernobyl (in June and July 1986) had already shown remarkably high levels in West Cumbria, near Sellafield (Institute for Terrestrial Ecology 1986). The fact that these measurements, which were taken from vegetation samples (Figure 1.2), did not altogether tally with the distribution as estimated from a combination of rainfall data during the crucial period while the radioactive cloud was over Britain, and models of rain-out of caesium from the atmosphere (Figure 1.3) also invited the question of whether a hidden factor, such as unacknowledged long-term Sellafield discharges, had created the differences (Smith and Clark 1986). This factor would

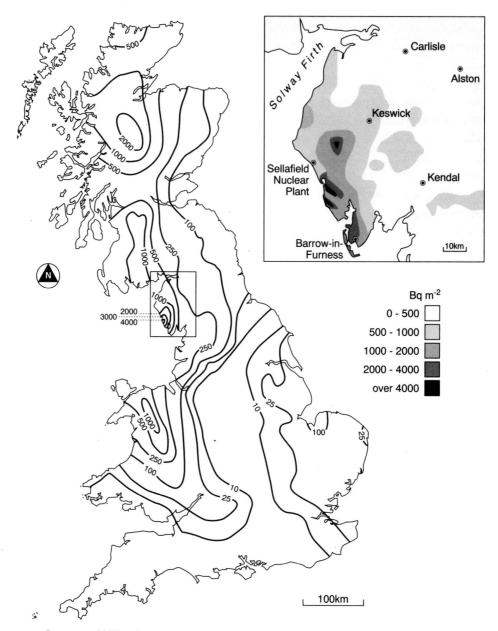

1.2 Contours of UK radioactive caesium contamination measured from vegetation, June–July 1986. The data are in units of Bq m⁻². (Source: Institute for Terrestrial Ecology (ITE))

be picked up by the vegetation-samples method, but not by the rainfall-data method.

In the manifest scientific confusion and inconsistency, it was as if the farmers had suddenly found an outlet for fears and suspicions that they had previously entertained, but felt unable to voice. Ironically it was radioactive contamination which

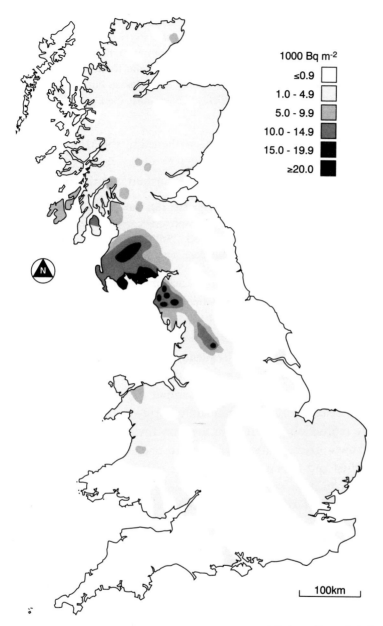

1000 Bq m⁻²

≤0.9

1.0 - 4.9

5.0 - 9.9

10.0 - 14.9

15.0 - 19.9

≥20.0

1.3 UK radioactive caesium levels estimated from rainfall data, Chernobyl cloud movement data, and models of caesium rain-out from the atmosphere

scientists confidently proclaimed was nothing to do with Sellafield-Windscale which gave the hill farmers their first embryonic voice about that local trouble-spot.

In our interviews, typical scepticism about the scientists' assertions of Sellafield's innocence was expressed as follows:

> There's another thing about this as well. We don't live far enough away from
> Sellafield. If there's anything about we are much more likely to get it from there! Most
> people think that around here. It all comes out in years to come; it never comes out at
> the time. Just look at these clusters of leukaemia all around these places. It's no
> coincidence. They talk about these things coming from Russia, but it's surely no
> coincidence that it's gathered around Sellafield. They must think everyone is
> completely stupid.[1]

These immediate local suspicions were only strengthened by the Ministry of Agriculture maps showing the restricted areas (Figure 1.1). Other farmers reinforced this
logic, as did experience of the continuing secrecy surrounding the 1957 fire.

> It still doesn't give anyone any confidence, the fact that they haven't released all the
> documents from Sellafield in 1957. I talk to people every week – they say this hasn't
> come from Russia! People say to me every week, 'Still restricted eh – that didn't come
> from Russia lad! Not with that lot on your doorstep.'

The scientific view was that the Chernobyl caesium depositions could be distinguished from the caesium in routine Sellafield emissions, 1957 fire emissions, or
1950s weapons testing fall-out, by the typical 'signature', in gamma-radiation energy
spectra, of the ratio of intensities of the isotopes caesium 137 and caesium 134 (each
isotope has a characteristic gamma-ray frequency or energy). The half-life of the
caesium 137 isotope is about thirty years, while that of the caesium 134 isotope is
less than one year, so the ratio of intensities of caesium 137 to caesium 134 increases
with time. A typical Sellafield sample (from fully burnt-up fuel, usually stored for
several years before reprocessing; or if from the 1957 fire, aged in the environment)
would therefore show a greater ratio (about ten to one) than a Chernobyl sample
consisting of fresh fuel and fission products (about two to one). Thus the deposits
were said scientifically to show the so-called Chernobyl fingerprint, making an analogy with a form of evidence which is never questioned in law.

   This scientific distinction, which exonerated Sellafield, was unequivocally asserted
at public meetings and lectures with virtually complete consensus from scientists
from the Ministry of Agriculture, Fisheries and Food (MAFF) and the other scientific organisations involved in the issue, at least for the first year or more of the
crisis. However, it too turned out to be less clear-cut than first claimed: it was later
admitted that only about 50% of the observed radioactive caesium was from Chernobyl, the rest being from 'other sources', including weapons testing fall-out and the
1957 Windscale fire (House of Commons Agriculture Select Committee, 1988; Farmers Weekly 1988). Nevertheless, at the time the difference in the fingerprints was
represented as a very clear-cut scientific distinction, with Sellafield for once in the
clear, and Chernobyl definitely to blame. Yet, although it was against their economic
interests to entertain thoughts of a longer-standing but neglected (or covered-up)
blight from Sellafield, and in the face of this solid scientific consensus, many hill

farmers were ready, at least in private, to implicate Sellafield. Their reasoning tells us a lot about the deeper cultural and social structures of expert credibility.

It was striking that when we asked farmers sceptical about the scientists' exoneration of Sellafield to explain their reasoning, many of them talked about the 1957 fire and the secrecy surrounding it. At first we did not see this as an answer to the question, but then we realised that it was – they were explaining the lack of credibility of the present scientific claim about the Sellafield–Chernobyl distinction as due to the untrustworthy way in which the experts and authorities had treated them over the 1957 fire, and the longer history of perceived misinformation surrounding Sellafield:

> Quite a lot of farmers around believe it's from Sellafield and not from Chernobyl at all. In 1957 it was a Ministry of Defence establishment – they kept things under wraps – and it was maybe much more serious than they gave out. Locals were drinking milk, which should probably never have been allowed – and memory lingers on.

The farmers thus embedded their reading of the present scientific claim about the isotope ratio distinction firmly within the context of the unpersuasive and untrustworthy nuclear institutional body language which had denigrated them for thirty years or more. Their definition of risk was in terms of the social relationships they experienced, as a historical process.

They had a range of further reasons supporting this dissident logic. The empirical evidence of the maps (Figure 1.1, and contrasts between Figures 1.2 and 1.3) was powerful as far as they were concerned; and official disclaimers were ridiculed with a heavy irony only evident in a personal interview, such as (referring to a MAFF scientist) 'she said she couldn't understand why the heaviest fallout from Chernobyl happened to fall around Sellafield'.

Thus the farmers gathered – and used – evidence which was drawn from science, including scientific inconsistencies on which the scientists themselves did not focus. They entered the scientific arena in this sense, redrew its boundaries, and, operating with different presuppositions and inference rules, also redrew its logical structures.

Other direct empirical connections were drawn which may not have made scientific sense, but which served to make a consistent explanatory picture to people who saw the science to be either politically manipulated or naively overconfident in its own certitude.

> Most farmers believe it's really from BNFL [Sellafield]. You'd have great difficulty convincing them otherwise. This area is a kind of crescent shape. If you're up on the tops [of the fells] on a winter's day you see the tops of the cooling towers, the steam rises up and hits the fells just below the tops. It might be sheer coincidence, but where the [radiation] hot spots are is just where that cloud of steam hits – anyone can see it if they look. You don't need to be a scientist or be very articulate but they've figured it out all right. I think there's been low-level fall-out ever since that place opened, and Chernobyl has gone on top of it.

Interestingly, the apparently unfounded notion that high deposits occur where 'the clouds' hit the fell sides is not unreasonable, because scientists themselves recognised the importance of intense 'occult deposits' of radio-isotopes direct from low-lying clouds and mists which are typical of the Lake District climate.

Other farmers seemed to be exercising a strong penchant for irony when they put into sceptical perspective the experts' claims about the 'coincidence' of Chernobyl deposition next to the local controversial nuclear site:

> When you look at the stations around here, I said it was like a magnet, it just drew it in! [Then, relaxing the irony] I still think it was here before. They [the experts] won't have it . . . We can't argue with them, but you can think your own ideas.

Often the justification for disbelieving the scientists on the Sellafield connection was simply that the same experts had very recently asserted, with similar confidence, first that there would be no effects of the Chernobyl cloud, and then that the restrictions which were imposed after all would be very short-lived. Since their self-confidence had been shown to be misplaced on those counts, why should they expect to be believed this time, especially when no admission of the earlier mistakes was forthcoming? The farmers scorned what they saw as the scientists' addiction to over-confidence and false certainty:

> My theory, which is probably as good as anyone else's is this: we don't know . . . They keep rushing to conclusions before the conclusion has been reached – you understand what I'm saying? They'd have been far better to keep their traps shut and wait.

And a National Farmers' Union local representative put it: 'We may be on the eve of a new age of enlightenment. When a scientist says he doesn't know, perhaps there's hope for the future!'

It is important to note that scientific credibility was influenced not only by the evidence which alternative logical presuppositions could select and render coherent, and not only by the prior intellectual mistakes, but by the way they were handled socially. This gave impetus to the alternative constructs.

The farmers also came into direct contact with the conduct of science on their farms, as hosts to a proliferation of monitoring, sampling, field analysis, and various other scientific activities. Again, they soon noted the inconsistency between the certainty pervading public scientific statements, and the uncertainties involved in actually attempting to create definitive scientific knowledge in such novel and open-ended circumstances. The experience of watching scientists decide where and how to take samples, of seeing the variability in readings over small distances, noticing the difficulty of obtaining a consistent standard for background levels, and of gradually becoming aware of the sheer number and variety of less controlled assumptions, judgements and negotiations that underpin scientific facts, corroded the wider credibility of official statements couched in a typical language of certainty and standardisation. By accident, as it were, the farmers entered the 'black-boxes' of constructed, 'naturally determined' science, and saw its indeterminacies for themselves (Latour 1987). Referring to the live monitoring of sheep by a government official which was obligatory, one farmer indicated how doubts set in:

> Last year we did 500 [sheep] in one day. We started at 10.30 and finished at about six. Another day we monitored quite a lot and about 13 or 14 of them failed. And he [the monitor] said, 'now we'll do them again' – and we got them down to three! It makes you wonder a bit . . . it made a difference . . . when you do a job like that you've got to hold it [the counter] on its backside, and sheep do jump about a bit.

These forms of reasoning were buttressed by further social evidence and judgement. There existed amongst the farmers a widespread model of the capture of science by institutions with their own manipulative political agendas. Such judgements were supported by empirical observations, such as the refusal of MAFF officials to allow an American television team to film the lively debate with affected farmers at a public meeting in February 1987. The TV team was preparing a five-country documentary on the international response to Chernobyl. The producer's acid comment as he departed that his team had received more open treatment in (*pre-glasnost*) Poland than in Britain – was widely quoted afterwards among the farmers (Williams and Wynne 1988).

The farmers drew similar conclusions from MAFF's response to their requests for pre-Chernobyl caesium data on the fell-top vegetation, soils, and sheep; they asked for these in order to test MAFF's assertion that there had been no significant contamination before Chernobyl. However, MAFF's reply was to refer first to an official document which contained only post-Chernobyl data (MAFF 1987), and then to data which included pre-1986 monitoring on the lowland coastal strip, but still no fell-top data. The farmers saw this as evidence that the authorities were trying to cover up – either that they did have data which showed *high* fell-top levels of caesium before Chernobyl, or that they had no data at all! If the former, they were guilty of straightforward lying and conspiracy to conceal a longer-standing contamination from Sellafield: if the latter, they were guilty of at least gross complacency and incompetence, but possibly also conspiracy to remain *deliberately* ignorant of the levels which prevailed before Chernobyl forced them to look. In addition, the 1957 fire had provided an ideal opportunity – apparently neglected – to have done the necessary research which would have avoided mistakes in the 1986 prediction:

> Going back to the 1957 fire, nobody really knows what that did, what effect it had on the land and that, because they never tested it . . . A lot of people have it in their minds that they [the UK authorities] were just waiting for something like this [Chernobyl] to blame.

This indicates a belief that the authorities had done secret research, had found high levels and had decided to cover up – waiting for the chance, which Chernobyl provided, to pass on the blame. It also encouraged the farmers to conclude that they and their families had been used as mere objects of scientific research.

In fact the question of whether the authorities had done previous research in the Cumbrian fells, and thus knew that the radioactive caesium contamination would last much longer, is extremely complex. What counts as 'previous research' is itself open to interpretive differences; some ecologists we interviewed said afterwards that they knew, and told the government at the time, but that they were ignored by the 'physicalist' ethos which dominated the official advisory mechanisms. This is the subject of further research. In evidence to the House of Commons Agriculture Select Committee in 1988, a local environmental group, Cumbrians Opposed to a Radioactive Environment, alleged cover-up, and also noted that the government's advisory body, the National Radiological Protection Board, had promulgated emergency reference levels for environmental radioactivity, only a month before Chernobyl, which completely omitted the central environmental medium and food chain in the Chernobyl emergency, namely sheep meat.

The feeling of being used for research rather than being assisted by scientific research was also reinforced by the offer which the authorities made to give people a whole body scan for radioactive contamination. This the farmers dismissed as useless information, being offered only so that the authorities could gather data, not to give people information they could use. Thus, whilst this offer was being made, the demand by the farmers for measurement of water-supply radioactive contamination was ignored, even though this was information they could have acted on. Again the scientists were exposed as ignorant or uninterested in local realities, this time imposing false assumptions about agency on local people. These modes of reasoning interlocked with other judgements which the farmers made of the controlling institutions from which scientific claims were seen to emanate. Thus another farmer related what he saw as deliberate official ignorance to Sellafield's denial of claims that the site caused leukaemia:

> The Department of Health could body monitor but they don't deliberately because if they did and found high readings then various ministries could one day be accused of irresponsibility in this regard. I think it self-evident that when BNFL [the Sellafield operator] were accused of being responsible for leukaemias they were quick to say 'what evidence is there?' I have been told that if I make an accusation that my granddaughter has got leukaemia in the future and I suggest it was due to Sellafield they will say to me 'what evidence have you?' It is a deliberate policy of government not to do this appropriate monitoring and testing so that they can protect themselves against an accusation of this nature. I would suggest we have another Christmas Island situation. The first such situation was at BNFL [it was then the Atomic Energy Authority] in 1957. Now we have Chernobyl Cumbria, Chernobyl Wales, South Scotland and Ireland ... When you have bottomless financial pits like Sellafield sponsoring this, that and the

other in order to blackmail local feeling, why could they not instead do something positive like supporting controlled experiments to answer all the questions that need to be answered?

Of course we can judge that these views were encouraged by probably unrealistic ideas about what can be expected of scientific knowledge in a situation such as the post-Chernobyl emergency. Even allowing for this factor however, the expressed attitudes reflect a rich supply of evidence to support a model, which lay people held, of the subordination of science to untrustworthy institutional and political interests, and of a deep flaw in the very nature of science which drives it towards unrealism, insensitivity to uncertainty and variability, and incapability of admitting its own limits. (These can be seen as contradictory models of science, but are better treated as rhetorical stances which deconstruct and delimit the authority of the social control which the science represented in the experience of the farmers.) Analysis of the logical structure of the farmers' responses to the scientific expertise indicates both a far greater open-endedness about scientific logical structures and its institutional and cultural forms than is usually recognised, and a greater need to acknowledge and negotiate these as a condition of science's social legitimation and uptake.

Credibility: the social dimension

The way in which the farmers' scepticism was expressed suggests that Chernobyl acted to release a large unrecognised and unexposed historical backlog of more private disbelief, mistrust, and alienation of local people from the authorities, which related to Sellafield; and which alienation had been quietly simmering and growing over the years as one experience of official perfidion led into another. This would also explain the apparently abrupt change in their position from acceptance to hostility: it was probably not nearly as abrupt as it may have seemed, because there was already a finely balanced 'private' ambivalence, and not by any means as complete and uncritical an acceptance of the site and its expert apologists as a superficial reading of public quietude might suggest (Wynne 1987).

However, the dimension of this issue which drew in the farmers, and on which they had the most confidence to judge the outside experts and to criticise them, was the fact that, this time, expert responses to the crisis constituted massive interventions, disruptions, and denigrations of their normal practices and livelihood. The administrative restrictions introduced by the government to prevent contaminated lamb from reaching the market were tantamount to large-scale social control and reorganisation, and denial of essential aspects of the farmers' social identity, to an extent that the outside experts and bureaucrats did not remotely recognise. The interventions required not only scientific understanding of the radioecology of caesium in this particular physical environment; they also required this to be integrated with knowledge of hill sheep-farming methods and decision-making processes, in what is a highly specialist and particular kind of farming. That is, the complex natural and social particulars of the situation of use of scientific knowledge needed to be understood and negotiated into an effective hybrid with the scientific knowledge.

Whereas the hill farmers were quite reserved in their scepticism towards the scientists on scientific matters, they were abrupt and outspoken about them when they saw the extent of the scientists' ignorance of hill-farming environments, practices, and decision-making. Even worse was the way that the outside experts did not recognise the value of the farmers' own expertise, nor see the need to integrate it with the science in order to manage the emergency properly. An example which ruined the experts' credibility with many farmers was the advice given to farmers to keep their lambs a little longer on cleaner valley pastures so as to allow high caesium levels gained on the fell tops to decrease. This ignored the locally taken-for-granted fact that hill farming in such areas is organised around a severe short supply of valley grass, which would, as one farmer put it, 'be reduced to a desert in days' (and with knock-on effects into future breeding) if it were not very carefully husbanded.

Naturally the farmers felt that their whole identity was under threat from outside interventions based upon what they saw as ignorant but arrogant experts who did not recognise what was the central currency of the farmers' social identity, namely their specialist hill-farming expertise. This expertise was not codified anywhere: it was passed down orally and by apprenticeship from one generation to the next, as a craft tradition, reinforced in the culture of the area. The impact of the scientists' hegemonistic cultural orientation on their general credibility showed itself repeatedly:

> There was the official who said he expected levels would go down when the sheep were being fed on imported foodstuffs, and he mentioned straw. I've never heard of a sheep that would even look at straw as fodder. When you hear things like that it makes your hair stand on end. You just wonder what the hell are these blokes talking about? When we hill men heard them say that we just said, what do this lot know about anything? If it wasn't so serious it would make you laugh.

Another derided the experts' ignorance of what were elementary facts of life to hill farmers:

> If you start fattening lambs and sell twenty, the next twenty get fat quicker, because you've got more grazing. But if you keep them all . . . [gesticulation of disaster]. But that's the problem with the ministry – trying to tell them those sort of things. That's where the job has fallen down a lot. They couldn't understand that you were going to sacrifice next year's lamb crop as well. They just couldn't grasp that!

Scientists and Ministry officials were often seen as indistinguishable; the most prominent officials explaining and defining official decisions were scientists. The organisational hierarchy within MAFF seemed to reflect such problems in that officials in the local Division at Carlisle, who did know and understand hill-farming culture, had no scientific standing, and so had little influence on the scientist policy advisers in Whitehall when they tried to act as a conduit for local farmers' knowledge. But there was also a deeper structural convergence between the forms of monopolistic scientific representation of the issues, and the forms of centralised administrative intervention and reorganisation of farming practices. The significant elements of scientific representation in this respect were:

- its artificial standardisation of variations in local physical environment, farming conditions, and practices (hence farmers); and
- its ethos of prediction and control, which engendered an exaggerated sense of certainty, and which conflicted sharply with the farmers' ethos of adaptation and acceptance of intrinsic lack of control.

These coincided with the centralised formal nature of the administrative interventions, which reduced the long-established individualism, informality, and flexibility of farm-management decision-making to an extension of bureaucracy. The farmers were quite familiar with uncertainty on several fronts and thus with adaptation to factors beyond their control. This deep cultural outlook – reflected in their intellectual frameworks as well as in their whole way of life – was simply incompatible with the scientific-bureaucratic cultural idiom of standardisation, formal and inflexible methods and procedures, and prediction and control.

These dimensions of incompatibility and lack of mutual credibility existed at a structural level which was deeper than that of evidence and information. They lay at the level of moral, or cultural recognition. Each side only recognised, even as possible evidence, claims expressed within its cultural style. Thus, for example, the scientists had an a priori credibility gap to overcome when they stated things so categorically and universally, before the substance of the statement was even reached. By the same token the farmers' expertise was not recognised because it was not formally organised in documentary, standardised, and control-oriented ways recognisable to scientific culture; and their later claims for compensation encountered the inflexible bureaucratic demand for formal documentation, dates, prices, numbers, proofs, and signatures in a way which was entirely alien to their own culture.

This sense of being ensnared by an alien and unrecognising combination of science and bureaucracy was neatly captured in two typical comments:

> They've been watching too much of 'One Man and his Dog' [a popular national television programme where shepherds compete in driving and penning sheep, under artificially simple conditions] . . . They think you just stand at the bottom of the fell and wave a handkerchief and the sheep come running.

Another, after a detailed explanation of complex differences between farming prac-

Table 1.1 *Lay criteria for judgement of science*

(i) Does the scientific knowledge *work*?   For example: predictions fail.
(ii) Do scientific *claims* pay attention to other available knowledge?   For example: scientists monitor sheep without paying attention to where they graze, whereas farmers know where on open fells they graze.
(iii) Does scientific *practice* pay attention to other available knowledge?   For example: when scientists devise and conduct field experiments which the farmers know will not work.
(iv) Is the *form* of the knowledge as well as the content recognisable?   For example: degrees of expressed certainty, standardisation, aggregation.
(v) Are scientists open to criticism?   For example: no recognition of other legitimate knowledges and expert actors; no admission of errors, omissions, or oversights.
(vi) What are the social/institutional affiliations of experts?   For example: imputed social/political biases and interests; historical track record of trustworthiness, openness.
(vii) What issue 'overspill' exists in lay experience?   For example: from Chernobyl to Windscale-Sellafield; lack of rational connection for scientists because institutional dimensions defined out a priori, but for lay people continuity depending on institutional models of agency and responsibility in decision and knowledge construction.

tices even within his own small valley, reflecting different microconditions, lamented: 'This is what they [the experts] can't understand. They think a farm is a farm and a ewe is a ewe. They think we just stamp them off a production line or something.'

Thus underlying overt clashes of knowledge, information, evidence, and belief were incompatible social and cultural structures, prescribed modes of social interaction. The scientific knowledge, in the aggregation and standardisation of data and parameters by which it was organised, also expressed commitments about the levels of political standardisation and control of the farmers, and in effect prescribed an alien cultural mode for them. It was far from simply 'information' to either use or misuse.

Thus the scientific perspective was just as socially grounded, conditional and value-laden as the other. Its credibility was influenced not so much by what it said directly and explicitly, as in the way it was institutionally and intellectually organised, including lack of recognition of its cultural and institutional biases – its own tacit social body language. As explained later, it suffered from its own lack of reflexivity.

Analysis of this credibility gap allows us to identify factors which affect the social credibility of science. These are summarised in Table 1.1 (see above), as criteria by which lay people rationally judge the credibility and boundaries of authority of expert knowledge. It is easier to understand the resilience of disputes over the authority of scientific knowledge when these several layers of the social and cultural framing of expert and lay discourses are recognised. They are structurally identical to the factors shaping the logics of dispute and development within science; it is just that in public situations the prior mechanisms of social closure are, by definition, less powerful.

This analysis suggests that reflexive recognition of its own conditionality is a pre-requisite for science's greater public legitimation and uptake; yet this requires more than an intellectual advance from science; it requires institutional reform of its modes of organisation, control, and social relations. This would involve, *inter alia*, recognition of new, socially extended peer groups legitimated to offer criticism of scientific bodies of knowledge from beyond the confines of the immediate exclusive specialist scientific peer group. The social definition of such extended peer groups would relate to the context of use of the scientific specialties concerned; and criticism would include explicit negotiation of the social criteria or epistemology of knowledge for the situation (Jasanoff 1990; Funtowicz and Ravetz 1992; Knorr-Cetina 1989). This approach to public understanding of science therefore underlines the point reflected in other sociological analyses of scientific knowledge, that the boundaries of the scientific and the social are social conventions, predefining relative authority in ways which may be inappropriate, and which are open to renegotiation (Jasanoff 1987; Star and Griesemer 1989). The practical process of developing that negotiation first requires recognition that existing approaches and discourse misrepresent this conventional character as if it were naturally determined.

Conclusions: lay reflexivity and social identities

A productive way of analysing the interactions between hill sheep-farmers and scientists in this case, is to see each social group attempting to express and defend its social identity. The farmers experience the scientists as denying, and thus threatening, their social identity by ignoring the farmers' specialist knowledge and farming practices, including their adaptive decision-making idiom. They also experience the scientists as engaged in a conspiracy with government against hill farmers, initially to deny any need for long-term restrictions and later to claim an innocent mistake in prediction – a combination of circumstances which influenced many farmers to make unfortunate decisions and to lose heavily as a result. Coming on top of the further hardships and external controls besetting the hill farmers in an area which is a tourist-dominated national park, their treatment by the scientists and bureaucrats after Chernobyl was almost the final straw in a baleful succession of blows to their cultural and social identity.

The scientists on the other hand were expressing and reproducing their intellectual-administrative framework of prediction, standardisation and control, in which uncertainties were 'naturally' deleted, and contextual objects, such as the farmers and their farms, were standardised and 'black-boxed' in ways consistent with this cultural idiom. Whatever private awareness they may or may not have had of the cultural limits and precommitments of their science, they successfully suppressed these.

These social identities were not completely predetermined and clear, nor were they immune to interactive experience and negotiation. My main point is that this dimension should be seen as the level from which explanation of lay responses to science is to be derived, and in which the factors and processes shaping credibility or 'understanding' can be identified.

The lay people in this case showed themselves to be more ready than the scientific experts to reflect upon the status of their own knowledge, and to relate it to that of others and to their own social identities. Thus, for example, the farmers implicitly recognised their social dependency upon the scientific experts as the certified public authorities on the issues, even if, as they indicated in interview, they held dissenting informal beliefs which they could defend along the lines described before. As one farmer put it: 'You can't argue with them because you don't know – if a doctor jabs you up the backside to cure your headache, you wouldn't argue with him, would you?', the suggestion being that when the expert tells you unbelievable things, you do not overtly argue, thereby inviting denigration. As another said: 'We can't argue with them, but you can think your own ideas. I still think it [the radioactive caesium] was here before.'

These more private beliefs were rarely displayed in public, and the farmers refused to confess to such overt dissent in media interviews. It was made clear to us that one reason for this was that the farmers identified socially with family, friends, and neighbours who were part of the Sellafield industrial workforce. They recognised their own indirect and sometimes direct social dependency upon the plant – not only neighbours, but also close relatives of the hill farmers work there. Thus, underlying and bounding their expressed mistrust of the authorities and experts, there was a countervailing deep sense of social solidarity and dependency – of social identification with material kinship, friendship, and community networks which needed to believe Sellafield was well controlled and its surrounding experts credible.

Thus social alienation and identification coexist in the same persons and communities, leaving deep ambivalence and apparent inconsistency in relevant beliefs and corresponding structures of 'understanding'. These can only be understood by reference to the multiplex, not necessarily coherent, dimensions of social identities expressed in interleaved social networks and experiences. Whilst trust is a key dimension of 'public understanding' and perceptions of risks, it should not be reified into an objective entity. It is a profoundly relational term, a function of the complex web of social relations and identities.

All this could be interpreted as yet another example of the lamentable inconsistency and impossible fickleness of lay beliefs. The conventional model of rationality would include a principle of cognitive consistency as measured against some canons of abstract logic. The 'rational' approach championed by modern scientific culture would assume inconsistency, imprecision, or ambivalence to be manifestations of intrinsic feebleness. However, we begin to see that such absolutist categories are actually moral or cultural stances, ones which assume control is an overriding value, yet without being able to recognise the cultural nature of this driving commitment. What is revealed in this case is a deeper and more complex consistency in public reasoning than that recognised by such simplistic models. In the real world people have to reconcile or adapt to living with contradictions which are not necessarily within their control to dissolve. Whereas the implicit moral imperative driving science is to reorganise and control the world so as to iron out contradiction and ambiguity, this is a moral prescription which may be legitimately rejected, or at least limited, by people. They may opt instead for a less dominatory, more flexible and adaptive relationship with their physical and social worlds. In this orientation, ambiguity and contradiction are not so much of a threat, because control and manipulation are not being sought or expected. This is no less legitimate a form of rationality than the scientists', and the 'public understanding of science' field needs to recognise this, and build upon it, in both research and in practice.

The advance from focusing on cognitive dimensions (often assumed public deficiencies) to trust and credibility is important. But closer examination in this case-study of the basis of trust and credibility falsifies the predominant analytical tendency to treat these as unambiguous, quasi-cognitive categories of belief or attitude which people supposedly choose to espouse or reject (Renn 1991; Jupp and Irwin 1989; Jupp 1989). My analysis suggests instead that 'credibility' and 'trust' are themselves analytical artefacts which represent underlying tacit processes of social identity negotiation, involving senses of involuntary dependency on some groups, and provisional or conditional identification with others, in an endemically fluid and incomplete historical process.

Thus what the hill farmers believed about the scientists and their assertions was rich in insights and refinement, on several levels beyond the one-dimensional reductionism of scientific logic alone. But this richness was also pervaded by an ambivalence reflecting their multiple and conflicting social networks and relations. It would have been easy to have marked them down as mere 'don't knows' in a more efficient attitude survey, even though this would have been a grotesque distortion of the true position. Indeed it is a serious challenge to large-scale survey methods to ask how they would have even identified these dimensions and whether they do not inevitably reinforce – normatively and conceptually – scientistic models of monovalent instrumental individuals.

Recognition of this multi-dimensional, even internally contradictory character of belief allows a more accurate perspective on the apparent fickleness of public responses to risks and scientific knowledge which is much lamented by authorities. If we assume that widely observed lack of public dissent to expert reassurances equals (voluntary espoused) public acceptance, then an apparently sudden shift to opposition and rejection seems capricious, irrational, and uncontrollably emotive. If, on the other hand, we recognise the tacit existing alienation and ambivalence often underlying surface quietude, we may see that what looks like a sudden shift of attitude, a 'betrayal', was nothing of the kind – it may have been only a very small shift in the complex balance of components of social identity which people are holding in tension with one another. This intrinsic instability of actors' 'loyalties' is something which is not fully recognised in Latour and Callon's theoretical vocabulary of enrolment and representation of actors by scientists, as they build intellectual-social empires by tying in those actors, appropriately defined, to their particular role in the edifice. Thus Callon's account of the 'betrayal' of the marine biologists by scallop fishermen of St Brieuc Bay who had seemed to have internalised the identity which the scientists had articulated for them, does not recognise the possible private ambivalence of the fishermen about their designated identity even before the 'betrayal', which may thus have been much less of a shift than it appears in Callon's otherwise superb account (Callon 1986).

Thus the cognitivist presumption that risks, or scientific knowledge, exist independently as an object for measurable public attitudes or beliefs, is left at least two steps behind. The first step is the recognition that the trustworthiness and credibility of the social institutions concerned are basic to people's definition of risks, or uptake of knowledge, and that this is reasonable, indeed unavoidable. Thus 'understanding' science is a function of experience, judgement and understanding of science's institutional forms as much as of its cognitive contents. However the second step is to recognise that trust and credibility are themselves analytically derivative of social relations and identity-negotiation; thus, like risk, they too should not be treated as if they have an objective existence which can be unambiguously measured and manipulated.

Having advanced the case for social identity as the more fundamental concept for explaining responses to science and risks however, it should be accepted that this term is itself not unproblematic. To claim that it offers more explanatory depth is not to claim it is empirically pure, coherent, and unambiguously identifiable.

The theoretical orientation of this chapter coincides with those perspectives which treat identities as intrinsically incomplete and open-ended, and as an endlessly revised narrative attempting to maintain provisional coherence across multiple social roles and reference groups (Hobsbawn and Ranger 1984; Giddens 1991; Shotter and Gergin 1988). Beliefs and values are functions of social relationships and patterns of moral and social identification. This stands in sharp contrast to the taken-for-granted (and hence rarely articulated) commitment underlying conventional approaches, in which values and beliefs are taken to be coherent, self-sufficient, and discrete entities, and where social identities are simply the aggregate of individual beliefs and values. In this perspective social interaction is recognised only as an instrumental device to maximise preferences and values, not as an activity with moral and social meaning in its own right (Burke 1991; Bailey 1968).

The case shows the unacknowledged reflexive capability of lay people in articulating responses to scientific expertise. They are able to reflect on and develop their own social position as part of a 'dependent' response in which they are supposed to enjoy no powers of independent critical rationality autonomous from 'proper' assimilation of scientific understanding. Indeed it is interesting that those who would be regarded as the representatives of traditional society showed this reflexive capability, whilst the putative representatives of enlightened modernity, namely the scientists, did not (Wynne 1993; Michael 1992). The scientists show no overt ability to reflect upon their own social positioning, that is upon the latent social models which their scientific interventions imposed on the farmers. Perhaps the distribution of reflexive capability (or impulse) is itself a contingent function of social relations of power.

It is not true to say that scientists are not reflexive, in that they do show a capacity to reflect upon the nature of their practice, its contingencies and limits. However, this may (for all social groups) be brought about only by criticism and a related sense of insecurity, rather than by any intrinsic qualities of self-criticism.[2] Thus the extent of such reflexivity in science is open to question, both in how deep it goes into examination of scientific founding commitments (hence identities) and in how openly such critical self-examination is expressed to other social groups, for example in public or policy debate. Such articulated self-criticism would display the uncertainties in scientific knowledge, and at the same time expose as negotiable science's definition and role in relation to other social groups. As I have suggested in this chapter, ambivalence is usually treated as intellectual feebleness – the antithesis of rationality and 'clear thinking'. But it may be a necessary corollary of a social commitment to disavowing control of others, in which case the remit of scientific rationality (as usually conceived) is seen in a radically different light (Harding 1986).

The issues and problems in public understanding of science thus cannot be divorced, as scientific bodies repeatedly assume they can, from the epistemological issues of the social purposes of knowledge, and what counts as 'sound knowledge' for different contexts. These in turn highlight questions about the institutions of science – its forms of ownership, control, and practice. To preclude these issues from public debate is to undermine the possibilities of an effective public uptake and culture of science, and in this sense self-consciously scientific institutions are often their own worst enemy.

The intellectual properties of reflexivity or its lack (or to put it another way, of the epistemological principles of science) are not independent of the institutional forms of science. Thus it becomes evident why the quality of its institutional forms – the organisation, control, and social relations – is not just an optional embellishment of science in public life, but an essential subject of critical social and cultural evaluation. It is a crucial missing part of the contemporary non-debate of science's social purchase and legitimacy.

ACKNOWLEDGEMENTS

Permission to publish an adapted version of a paper previously published in the journal *Public Understanding of Science* 3,3 (1992), is gratefully acknowledged. The work was supported by the UK ESRC. I am also grateful to colleagues and friends under the Science Policy Support Group network of research on public understanding of science, for rewarding discussion and moral support, especially to Peter and Jean Williams, John Wakeford, Mike Michael, Alan Irwin, Rosemary McKechnie, and Frances Price.

NOTES

1. Williams, P., and Williams, J. The quotes are from transcripts of interviews, which were taped and then transcribed in abridged form to record elements of relevance to this study. Over fifty interviews were conducted with affected farmers, farmers' wives, MAFF officials, scientists, farmers' representatives, and others. Each interview lasted between one and two hours; several repeat visits were made, allowing some observation of changing beliefs. The interviews were mostly conducted by Peter and Jean Williams, accompanied by the author on about twelve occasions. Public meetings and markets were also attended and observed. All the quotes in the text are verbatim quotes from interview tapes.

2. This is a distinction recognisable in the approaches to reflexivity of Beck and Giddens. Whereas Giddens (1991) tends to assume that reflexivity is an intrinsic self-transforming property of science, and hence of modernity as influenced by scientific culture, Beck (1992) recognises the social basis of scientific knowledge which means that moral and cultural commitments inevitably become identified with scientific knowledge. Thus reflexivity is in Beck's view the result of political criticism and contradiction rather than an inherent institutional trait. Although I have other criticisms of Beck's perspective (Wynne 1996), this chapter is on this aspect much closer to Beck than to Giddens.

REFERENCES

Arnold, L., 1992, *Windscale 1957: the anatomy of a nuclear accident* (London, New York: Macmillan).

Bailey, F. G., 1968, 'A peasant view of the bad life. *Peasants and Peasant Society*, edited by T. Shanin (Harmondsworth: Penguin).

Beck, U., 1992, *Risk Society: towards a new modernity* (London, Newbury Park, New Delhi: Sage).

Beresford, N. A., Howard, B., and Horrill, A. D., 1989, 'The effect of treating pastures with bentonite on the transfer of ^{137}Cs from grazed herbage to sheep', *Journal of Environmental Radioactivity* 9, 112–22.

British Nuclear Fuels plc, 1987, *Annual Reports on Environmental Monitoring* (London: BNF plc).

Burke, P., 1991, 'We the people: popular culture and popular identity in Modern Europe', *Modernity and Identity*, edited by S. Lash and J. Friedman (Oxford, Cambridge, Mass.: Blackwell) pp.293–310.

Callon, M., 1986, 'Some elements of a sociology of translation: domestication of the scallops and the fishermen of St. Brieuc Bay', *Power, Action and Belief*, edited by J. Law, Sociological Reviews Monographs, no. 32 (Keele University Press) pp.196–233.

Clark, M. J., and Smith, F. B., 1988, 'Wet and dry deposition of Chernobyl releases', *Nature*, 332, 245–9.

Durant, John R., Evans, G. A., and Thomas, G. P., 1989, 'The public understanding of Science', *Nature*, 340, 6 July.

Farmers Weekly, 4 and 11 March 1988, 'Chernobyl I and Chernobyl II' (London).

Friends of the Earth, 1987, *Fallout over Chernobyl: a review of the official radiation monitoring programme in the UK*, edited by P. Green and P. Daly (London: Friends of the Earth).

Funtowicz, S., and Ravetz, J., 1992, 'Three types of risk assessment and the emergence of post-normal science', *Social Theories of Risk*, edited by S. Krimsky and D. Golding (New York: Praeger).

Gardner, M. J., Snee, M. P., Hall, A. J., Powell, C. A., Downes, S., and Ferrell, J. D., 1990, 'Results of case-control study of leukaemia and lymphoma among young people near Sellafield nuclear plant in West Cumbria', *British Medical Journal*, 300, 423–5.

Giddens, A., 1991, *Modernity and Self-Identity* (London: Polity Press).

Harding, S., 1986, *The Science Question in Feminism* (Milton Keynes: Open University Press).

Hobsbawn, E. J., and Ranger, T., 1984, *Invention of Tradition* (Cambridge University Press).

House of Commons Agriculture Select Committee, 1988, *Chernobyl: the Government's Response* (London: HMSO).

House of Commons Agriculture Select Committee, 1988, *Chernobyl: the Government's Response*, vol. 2: Minutes of Evidence (London: HMSO).

Howard, B. J., and Beresford, N. A., 1989, 'Chernobyl radiocaesium in an upland farm ecosystem', *British Veterinary Journal*, 145, 212–24.

Institute for Terrestrial Ecology, 1986, Merlewood, South Cumbria. Map first published in *The Guardian* newspaper, 25 July.

Jasanoff, S., 1987, 'Contested boundaries in policy-relevant science', *Social Studies of Science*, 17, 2, 195–230.

Jasanoff, S., 1990, *The Fifth Branch: Science Advisers as Policymakers* (Cambridge, Mass.: Harvard University Press).

Jupp, A., 1988, 'The provision of public information on major hazards', unpublished M.Sc. thesis, Manchester University.

Jupp, A., and Irwin, A., 1989, 'Emergency response and the provision of public information under CIMAH – a community case study', *Disaster Management*, 1, 4, 33–8.

Knorr-Cetina, K., 1989, 'Epistemic cultures: forms of reason in science', *History of Political Economy* 23, 105–22.

Latour, B., 1987, *Science in Action* (London: Open University Press).

Livens, F. R., and Loveland, P. J., 1988, 'The influence of soil properties on the environmental mobility of caesium in Cumbria', *Soil Use and Management* 4, 3, 69–75.

Macgill, S., 1987, *The Politics of Anxiety: Sellafield's cancer-link controversy* (London: Pion Press).

MAFF's Annual Terrestrial Radioactivity Monitoring Programme, TRAMP, No. 1, 1987 (London: MAFF).

McSorley, J., 1990, *Living in the Shadow: the story of the people of Sellafield* (London, Sydney: Pan Books).

Michael, M., 1992, 'Lay discourses of science: Science-in-general, science-in-particular, and self', *Science, Technology & Human Values* 17, 313–33.

Ministry of Agriculture, Fisheries and Food, 1987, *Radionuclide Levels in Food, Animals and Agricultural Products* (London: HMSO).

Renn, O., 1991, 'Risk communication and the social amplification of risk', *Communicating Risks to the Public: international perspectives*, edited by R. E. Kasperson and P. J. M. Stallen (Dordrecht, Lancaster: Kluwer Academic) pp.287–324.

Shotter, J., and Gergin, K., 1988, *Texts of Identity* (London: Sage).

Slovic, P., 1992, Perceptions of Risk: reflections on the psychometric paradigm in *Social Theories of Risk* edited by S. Krimsky and D. Golding.

Smith, F. B., and Clark, M., 1986, 'Deposition of radionuclides from the Chernobyl cloud', *Nature*, 322, 690–1.

Star, S. L., and Griesemer, J., 1989, 'Institutional ecology; translations and boundary objects: amateurs and professionals in Berkeley's Museum of Vertebrate Zoology, 1907–1939', *Social Studies of Science*, 19, 3, 387–420.

Wynne, B., 1980, Technology, risk, and participation: The social treatment of uncertainty', *Society, Technology and Risk*, edited by J. Conrad (London: Academic Press) 167–202.

Wynne, B., 1982, *Rationality and Ritual: the Windscale inquiry and nuclear decisions in Britain* (Chalfont St. Giles: British Society for the History of Science).

Wynne, B., 1987, 'Risk perception, decision analysis and the public acceptance problem', *Risk Management and Hazardous Wastes: implementation and the dialectics of credibility* (London, Berlin, New York: Springer), chapter 11.

Wynne, B., 1989, 'Sheepfarming after Chernobyl: a case study in communicating scientific information', *Environment*, 31, 2, 10–15, 33–9. Reprinted in Bradby, H., 1991, *Dirty Words: writings on the history and culture of pollution* (London: Earthscan) pp.139–60.

Wynne, B., 1991, 'Public understanding and the management of science', *The Management of Science*, edited by D. Hague (London, New York: Macmillan).

Wynne, B., 1992, Risk and Social learning: Reification to engagement', *Social Theories of Risk*, edited by S. Krimsky and D. Golding, chapter 12.

Wynne, B., 1993, 'Public uptake of science: A case for institutional reflexivity', *Public understanding of Science*, 2, 321–37.

Wynne, B., 1996, 'May the sheep safely graze? A reflexive view of the expert–lay knowledge divide', *Risk, Environment and Modernity: towards a new ecology*, edited by S. Lash, B. Szerszynski, and B. Wynne (London: Sage).

2 Science and Hell's kitchen: the local understanding of hazard issues

ALAN IRWIN, ALISON DALE, AND DENIS SMITH

> You'd be more worried if you went round. It would frighten you to bloody death. You'd just see Hell's kitchen – you wouldn't know how they controlled everything.
>
> *(Manchester resident discussing possibility of a visit to the local chemical works)*

In this chapter, the focus is on environmental threats – one of the most topical and pressing areas of public debate and controversy involving science. Rather than considering environmental issues as they relate to global or national concerns, we will focus on immediate questions of pollution and hazard as they affect specific local communities. Whilst Steven Yearley tackles some of these questions in Chapter Eight with regard to environmental organisations, our attention will be concentrated on community residents who have a more 'everyday' approach to living in a hazardous environment. Given the present concern over environmental issues, what sources of information exist and how useful do they appear? In what ways do people relate to and 'make sense' of those sources of expertise? Going further, we will also consider the relationship between the provision of technical information and the social context within which that information is developed, disseminated, and received.

The central theme of this chapter is, therefore, that of 'citizens' and 'sources' set against matters of environmental concern. In this we are approaching a topic which recurs throughout this book – albeit, as we will see, from a distinctly 'local' and community-oriented perspective. We are also selecting a theme which was identified by the Royal Society as a gap in current knowledge: 'We therefore also recommend that the sources from which individuals obtain their understanding of science be actively investigated.'[1] However, whilst it is indeed important that we address such practical questions as *where* citizens turn for advice and information about technical matters or what *motivates* them to do so, our discussion of 'citizens and sources' also needs to take account of a number of more theoretical factors. At this stage, three particular aspects of the perspective adopted in this chapter need to be stressed.

Firstly, it seems fundamental to an appreciation of the public understanding of – in this case – hazard issues that full appreciation is accorded to the pre-existing knowledges and information networks possessed by citizens. The conventional approach to these questions is to assume a crude 'deficit' (or *tabula rasa*) model of the science–citizen relationship. The emphasis by the Royal Society and others on science and its dissemination can lead to the adoption of a 'top–down' model where only certain forms of knowledge are seen as privileged and legitimate. From this high-science perspective, the central issue is the public failure to understand science. Rather than assuming this perspective, our intention here is to portray the provision of technical information as part of an *interaction* with other forms of knowledge, understanding, and communication. More particularly, we wish to highlight those context-specific (or 'local') knowledges which are shared by particular groups of citizens. To what extent, for example, do such local knowledges match with, contradict, or simply ignore the 'official' technical statements emerging from industrial sources? Rather than simply taking for granted from the outset a 'public failure', the intention in this chapter is to consider just how valid the Royal Society model really is. Hopefully, we can also shed some light on *why* differing public reactions to science occur.

Secondly, and equally importantly, at the same time as appreciating the diversity and range of *citizen* understandings we should also be alert to the variety of *technical* expertises which are made available to citizens. Thus, rather than seeing 'science' as an undifferentiated mass, it is important to explore the *kinds* of technical information which are offered by different social groups (for example, local industry, councillors, the town hall) in different social situations. In addition to having an effect on the *selection* of information for dissemination (who should be told what and when), it is possible that citizen demands will also have an effect in *shaping* the technical knowledge which is developed (for example, citizen demands for more information concerning the health effects of low-level pollution can actually stimulate new lines of scientific research if only for defensive reasons on the part of industry). Rather than portraying citizens as the 'problem', therefore, we need to maintain a critical perspective on the selective development and dissemination of technical information. In suggesting this focus, we should also be alert to the possibility that different technical knowledges will embody different institutional and social commitments.

Thirdly, and as a consequence of the above two points, it is important to recognise that science as such often 'disappears' from view within local discourses – in this case, concerning environmental threat. At one level, therefore, the communications and responses discussed in this chapter make little use of the language of science, nor do scientific experts play a central role in community discussions of the pollution issue. Nevertheless, implicit models of technical expertise are deployed in, for example, the messages of reassurance from industrial officials to local residents concerning the safety of the chemical works on their doorstep. On a specific occasion, this meant that one of the companies used quantitative data about workplace safety in order to suggest that *public* safety was improving (but without drawing attention to the differences in these categories). Science and technology serve in this loaded social context as a general ideological support for current social and institutional practices. In particular, scientific expertise may be used as an underpinning for the suggestion that certain institutions can be *trusted* by public groups. However, since the companies in question are not generally exposed to technical challenge, this important legitimatory function remains largely implicit. Nevertheless, residents are sceptical of 'experts' and their relationship to industrial interests. Science, therefore, plays an important role within community risk discussions – but at a level removed from that suggested by the Royal Society.

The points made so far in this introductory section all suggest the significance of the social and cultural context in which knowledges and understandings (whether explicitly 'technical' or not) are developed and communicated. As will be argued throughout this chapter, it is impossible to comprehend the 'public understanding of science' without also comprehending the social and cultural processes through which public knowledge is communicated and generated. Thus, we need to be sensitive to the possibility that different forms of 'knowledge relation' may exist in different social settings. In this chapter, that setting is one of 'community responses to pollution' which may distinguish it from the other cases in this book.

In tackling these issues here, we will begin from the assumption that – far from being 'information poor' – groups of citizens actually have access to a range of cultural resources: 'A culture is common meanings, the product of a whole people, and offered individual meanings, the product of a man's whole committed personal and social experience.'[2] The issue to which we will return throughout this chapter concerns the place of science within this network of 'common meanings'.

In order to pursue the 'citizens and sources' issues as redefined above, our presentation will focus on a detailed study of two specific localities within Greater Manchester (in the north west of England). Both of these (i.e. the Clayton/Beswick area of east Manchester and Eccles within the city of Salford) can be characterised as highly urban and also industrial in character. Each possesses a large amount of hazardous industry close to major housing areas.

This housing consists of old terraced properties as well as more modern council accommodation – the latter a cause of particular concern due to its relatively recent construction close to the chemical industry. Indeed, one estate in Eccles was built during the 1970s reportedly against the advice of the Health and Safety Executive (HSE) on the grounds of major hazard risk. However, the local authority was keen to utilise the land in order to rehouse those previously living in an area earmarked for slum clearance – necessitating a relocation of some four miles which one resident described as 'like coming to Australia'. Each area is criss-crossed with a number of main roads carrying both commuter and industrial traffic (including chemical tankers to and from the works). While certain pockets of middle-class housing can be found in the two areas, the majority of residents can be reasonably unproblematically classified as working class.

Having selected these two areas – largely on the grounds of physical accessibility, researchability in terms of local contacts, and the inherent interest for us presented by the two communities (which at first sight displayed both fascinating similarities and contrasts) – we then focused on a number of high-salience hazard issues in order to explore citizen information-seeking strategies and communication processes. Thus, our approach has been to capture a specifically *local* dimension to the relationship between citizens and sources and – within this – to select highly visible topics where some efforts might already have been made to seek out information and advice. The topics selected were pollution and major accident hazards from nearby chemical plants, road traffic safety and (as a less community-oriented contrast) health and safety at work.

Our main method of exploring these issues was through a small-scale doorstep questionnaire followed by a series of semi-structured interviews. In what follows, we will draw upon both questionnaire and interview data and concentrate on responses to the hazards of the local chemical industry and on the information sources that exist. The *methodology* used in our study – and especially the use of survey data – marks it apart from the other projects in this collection which typically draw upon more qualitative and ethnographic data. However, in this case it seemed valuable to *combine* qualitative and quantitative methods in order – at least potentially – to contrast the picture of everyday 'common meaning' which emerged from each. Of course, both our research methods focused on well-defined geographical locations rather than seeking to present a de-contextualised view of 'public understanding'. In that way, the methodology employed here differs radically from attempts at national surveys of attitudes and understandings.

Citizens, sources, and local environmental threats

One immediately striking aspect of our investigation was the indicated level of concern in the two selected areas over hazard issues. Thus, comparable questionnaire surveys have typically indicated that such issues figure relatively low in comparison to other social issues such as unemployment, violence, and housing conditions. However, replies to our question 'Is there anything about living in this area that you dislike or that worries you?' suggested that not only was there a high level of general concern about the area (with only 31.8% replying 'no' to that question), but also that hazard issues figured prominently as part of that concern. As one approximate indicator of this, we present in Table 2.1 the percentage of questionnaire respondents

Table 2.1 *Comparison of the two areas – issues of concern (prompted)*

	ECCLES		CLAYTON	
	No.	%	No.	%
State of the health service	105	56.5	62	36.3
Crime and violence	151	81.2	130	76
Factory accidents or pollution from factories which might affect the neighbourhood	86	46.2	98	57.3
Education problems	47	25.3	28	16.4
Risks of traffic accidents	56	30.1	52	30.4
Inflation	105	56.5	69	40.4
Health and safety at work	22	11.8	19	11.1
Unemployment	93	50	94	55
Housing conditions	52	28	41	24

in the areas who (when prompted on each) considered the above issues to be a matter of concern. The percentages are of total responses from that area.

Whilst it will be argued throughout this chapter that questionnaire data needs to be generated, presented, and interpreted with great care lest it conceal underlying social and cultural processes, Table 2.1 suggests a striking picture of concern over factory accidents/pollution. This concern emerges all the more strongly given the areas in question which have been greatly affected by unemployment and urban decay. Especially in Clayton, pollution and explosion as a consequence of the chemical industry represent a major topic of local anxiety. Less formal interviews in both Eccles and Clayton reinforce this picture with constant references to noise, smell, and atmospheric pollution ('there's a corn-flakes smell and a blue haze, every chimney is on the go at night' – Eccles resident), and also to the risks of explosion ('If anything did go up it would take everything in a seven-mile radius. That's what's said generally amongst local people' – Eccles resident), as well as general health problems ('This place is top of the league for chest complaints' – Eccles resident).

Undoubtedly, therefore, environmental questions figure prominently amongst the everyday concerns of these two communities. These concerns seem to have been triggered and reinforced by a history of relatively minor pollution incidents and persistent local complaints. However, it is important to place this concern with hazard in the context of other community concerns – especially those surrounding unemployment. We take just one exchange between two Clayton residents as an illustration of this:

R1– I would say most people around here worry about the Aniline.

R2– If Clayton Aniline shut down it would be a bloody ghost town around here.

This theme of hazard issues balancing against employment concerns reflects well the reality of two areas which have been subject to a considerable amount of de-industrialisation in recent years. The perceived tension between the general argument that 'industry like this shouldn't be on our doorstep', and also that 'what this area needs is more jobs', forms a regular basis for discussion. Already, we see something of the dynamic local context within which different information sources will be interpreted – and also generated.

If we move beyond this specific tension, concern over hazard is portrayed more broadly as one part of the pressures (and pleasures) of everyday life in a particular community – a point made in different ways by the above two Clayton residents:

R2– I like the people, that's what I like about the place. No thuggery. I like my house. There's a factory outside, but so what? I wouldn't want to move.

R1– . . . but you wouldn't want to move from your work. We stayed here for the work, the schools, house prices are relatively cheap. I'm happy with the area. It's not brilliant for shops but that's no problem. It's that place [the local chemical plant] that worries me. They've got a duty to tell people things, but they've neglected it. We're living on a time bomb. I bet there's more going on than we know.

The above quotations suggest the variety of local responses to environmental issues, but also indicate that these responses form part of a wider assessment of the area and what it means to live there. Whilst, from an outsider's perspective, hazard issues can be separated off from other concerns, for a member of the community the picture is more complex – a point which was made especially well in a group interview in Eccles. At the end of a one-and-a-half hour session, one of us attempted (somewhat clumsily) to conclude a particularly lyrical discussion of environmental issues only to be interrupted with a plea of 'what are we going to do about his rent increase?'. Although this caused some embarrassment to other participants, the underlying message was quite clear – this issue is just one part of everyday life, do not take it out of context or grant it unreal prominence.

What is true of hazard issues is, we will argue, also true of science and issues of 'public understanding'; it is highly artificial to filter out just this factor from its larger social and cultural context. Whilst, from the perspective of the scientific community, science 'disappears' within everyday life; from the perspective of local people, science has no necessarily special or privileged status alongside such routine concerns as unemployment, rent increases, or factory pollution. Issues such as 'knowledge' or 'hazard' thus form part of a wider pattern of everyday social relations.

These general points about the context to hazard and information issues take on full significance as we now address specific questions of 'citizens and knowledge sources'. If we return to our formal survey, one question put to those who had expressed concern (either prompted or unprompted) over chemical hazards was

Table 2.2 *'Have you ever been given, or tried yourself to get, information or advice about the local chemical industry from . . .?'*

	No.	%
Chemical company	52	28.3
Community group	8	4.3
Emergency services	3	1.6
Local authority	9	4.9
Citizens Advice Bureau (CAB)	2	1.1
Local councillor	3	1.6
MP	1	0.5
HSE	0	
Elsewhere	4	2.2

whether they had been given, or tried themselves to obtain, information or advice from a number of sources. Responses to this are summarised in Table 2.2.

Three aspects of this table deserve immediate comment. First of all, there is the strikingly small proportion of those concerned about this issue who consider themselves to have either sought or received information. A similar situation occurs with road traffic accidents and workplace safety. The second striking feature of Table 2.2 is the relatively high significance given to local industry as an information source. This point accords well with the Royal Society's identification of industry as a major source of technical guidance about their activities. Certainly, a strong feeling in our study areas was that, since industry was likely to be the original source of all information about risks and pollution, then it was probably just as well to go there directly ('Cos it's straight from the horse's mouth'). Thirdly, a number of local people pointed out that not only has local industry the relevant information, but also that it is 'convenient and close' *and* (very importantly) that 'if enough people went they might do something'. At this point, we begin to see the linkage between 'information seeking' and 'action seeking'.

These three points concerning the *number* of requests, their *focus* on industry and their underlying *motivation* all need to be interpreted within the social context of these two areas. One important dimension of this can be illustrated with reference to a further part of our questionnaire. Table 2.3 represents the formal responses to our question concerning the *trustworthiness* of various possible sources. As can be seen, an interesting contrast emerges with Table 2.2.

Table 2.3 provides an intriguing snapshot of local opinion in our two study areas. More importantly, it also implies something of the *critical evaluation* made by citizens of potential information sources. Despite the regular assumption by those in scientific institutions that 'science' is a straightforward and unitary category (so that the

Table 2.3 *'There are various places that could provide you with information or advice about the local chemical industry. How trustworthy do you think the following would be as sources of information on such matters?'*

	Very trust-worthy %	Trust-worthy %	Untrust-worthy %	Very untrust-worthy %	Don't know %
Local chemical companies	9 (4.9)	41 (22.3)	71 (38.6)	34 (18.5)	29 (15.8)
Local community groups	20 (10.9)	99 (53.8)	8 (4.3)	0	57 (31)
Police	15 (8.2)	97 (52.7)	19 (10.3)	2 (1.1)	51 (27.7)
Fire service	68 (37)	91 (49.5)	2 (1.1)	0	23 (12.5)
Local authority	8 (4.3)	73 (39.7)	30 (16.3)	7 (3.8)	66 (35.9)
CAB	32 (17.4)	104 (56.5)	6 (3.3)	0	42 (22.8)
Local councillors	6 (3.3)	55 (29.9)	40 (21.7)	12 (6.5)	71(38.6)
Local MP	10 (5.4)	53 (28.8)	23 (12.5)	8 (4.3)	90 (48.9)
Health and Safety Executive	36 (19.6)	81 (44)	9 (4.9)	0	58 (31.5)

problem of 'public understanding of science' is seen as one of getting *more* science to the public rather than of assessing the *kinds* of science which might be made available), the picture here is of local citizens discriminating between different 'knowledges' according to their source and the perceived social credibility of that source. However, 'trustworthiness' must be balanced against some assessment of the capability of that source in terms of citizen needs for practical action as well as for information.

In order to illustrate this complex pattern of local assessments we can offer the example of the fire service which emerges from Table 2.3 as the most trustworthy of all – largely on the grounds that the service enjoys a high social standing ('They're great men'). Whilst no one was prepared to disagree with this assessment, there was, however, a general feeling that, whilst they would certainly be honest and truthful ('They'd have to be – they've got responsibility to the community'), they would not be in a position to serve as an independent information source ('They could only tell you just what the factory say they can tell the public').

This assessment of the trustworthiness and technical capability of the fire service can be contrasted with, for example, that of 'local community groups' which emerged from Table 2.3 as relatively trustworthy – albeit with a reasonable proportion of 'don't knows'. Comments here were more cautious than those regarding the fire service ('Depending if they were involved in the local chemical industry') and unsure as to whether they have the necessary resources ('They wouldn't know much about it'). Thus, although seen as potentially having an important role to play (especially since this source was widely portrayed as sympathetic and helpful), reservations were expressed about the kinds of advice which such groups might make available.

All of these points take on special weight if we now turn our attention to the social organisation which is widely seen as the most important in these matters – the chemical companies. As Table 2.2 suggested, this is the most significant source of 'technical' information and advice for local people. However, perhaps the most striking feature of Table 2.3 is that, despite this significance as an actual source, the 'untrustworthy' rating given to local chemical companies is the highest of all. Whilst it would be unfair to suggest that *all* local citizens share this scepticism – and it must be noted that more people in Eccles than in Clayton consider the industry to be 'trustworthy' – it can be stated that debate over the trustworthiness of local industry forms a central part of social discourse over these matters. Once again, therefore, we see the critical climate within which technical information is received and, in particular, a local awareness of the importance of the information's *source*.

This critical treatment of information sources in terms of their perceived trust-worthiness and technical/practical capability represents an important part of the local knowledge of life in these areas. Certainly, for example, any perceived disparity between reassurances offered and subsequent (or previous) pollution incidents will be seized upon by the community as a measure of untrustworthiness. Thus, 'information' does not exist in a social vacuum, but is weighed in terms of previous experience and cultural evaluations. Such local knowledges will consist of a record of previous incidents together with a variety of other factors such as an economic assessment of local industry, the popularity of senior management, shared knowledge of production processes and final products, the attitudes and opinions of friends and neighbours, and the generalised outcome of simply living with a hazard site over a period of time. Of course, compared to the formalised structures of science, this knowledge will typically appear unsystematic and individual – being the outcome of a process of bricolage[3] rather than training. The knowledge structure will also, as we have argued, view hazard and information issues as one part of a larger set of concerns about the area and everyday life. However, none of these points imply that local understandings are inadequate or deficient in comparison to formal, scientific under-standing. On the contrary, they may well represent a more robust and well-tested body of advice, information, and practical assistance than any new or externally generated piece of technical evidence.

Some sense of the nature of this local discourse can be gleaned from comments made during the questionnaire from those assessing the trustworthiness of 'local chemical companies'. Comments are listed according to questionnaire response in order to suggest the underlying pattern of assessment;

Local chemical companies –

VERY TRUSTWORTHY 'Wouldn't hide anything'

TRUSTWORTHY 'They'd have to tell the truth but they'd hold back a bit'
'Perhaps, but probably just tell you what they want you to know'

UNTRUSTWORTHY 'Well, I reckon they'd be a bit careful what they told you'
'I think they'd only tell you a load of blinding science to shut you up'
'They wouldn't give you the full view – they've got their jobs to think of'

VERY UNTRUSTWORTHY 'Wouldn't believe a word they said'
'They'd only tell you something to put your mind at rest'

DON'T KNOW 'They would say one thing and mean another'
'They would just let you know what they wanted you to know'

What is striking about the above comments is the extent to which – almost regardless of the specific 'trustworthy' categorisation made by the respondent – they fit a pattern of scepticism regarding local industry as a source of information. As suggested by the above comments, this pattern seems as true of those who are sympathetic to the chemical companies as of those who are more negative. To a high degree, local citizens see industry as self-interested and as offering a very partial account of environmental hazards from their activities. For example, one local science teacher who saw the local chemical works as just 'doing their job' and as 'generally quite responsible' felt that: 'the company would be the best place to go, but they would only tell you certain things ... They'd be capable of pulling the wool over [my] eyes' (Eccles resident). Equally – and somewhat problematically for local industry – enhanced efforts at information-giving can actually provoke *greater* suspicion; 'it's only something for the company's benefit . . .'; 'They must be making a lot of money to try so hard with us' (Clayton residents).

On the other hand, as we have suggested above, industry is seen as the only party who can actually *do* anything about hazard reduction and, therefore, it is to them that complaints and requests for information need to be addressed. However, the fact that such a source will be consulted does not mean that its response will simply be accepted at face value. Instead, the local culture encourages a critical evaluation of new information in terms of the social assessment of the trustworthiness (and degree of self-interestedness) of its source. Needless to say, this generalised social response incorporates the judgements of *science* as well as other forms of output from industry. Thus, the reference above to 'blinding science just to shut you up' reveals much about the contextualised judgement regarding 'scientific' dissemination. The model of science possessed by these residents (i.e. as straightforwardly reflecting dominant social interests) seems far again from the notion of 'pure' science apparently held by groups such as the Royal Society.

Of course, this critical evaluation does not apply solely to local industry – all other sources are likely to (and, indeed, do) undergo similar scrutiny. One regular focus of interview discussion concerned the possibilities of 'independent' advice and assistance, i.e. emanating from sources which would have no 'stake' in local developments. In particular, such advice was seen as potentially valuable in offering an alternative to industry-based information: 'a letter from a scientist – that would be OK ... people would take notice. I'm not saying I'd swallow everything. I'd listen then make up my own mind. There's no point in listening to the company. Better to listen to the scientists' (Clayton resident).

Whilst some residents advocated the use of 'outsiders' (for example, non-partial scientists) as an unbiased source of information, there was also an inevitable degree of scepticism surrounding the neutrality of such individuals: 'The outside person would tell you it's dangerous, but the company might co-opt them. The company might give them money for their project, it might have got to them' (Clayton resident).

The plea for totally 'independent expertise' may, therefore, in practice be unmatchable. It also seems to suggest that there *is* seen to be such a concept as 'pure' science – but *not* within the everyday context as defined by local residents. This view of science as worthy and important but 'out there' contrasts sharply with the assessment of locally provided information which is co-optable and linked to its particular source – and, therefore, not considered to be 'scientific'. Such an idealised notion of 'science' as opposed to 'what industry tells us' reinforces the 'disappearance' and perceived irrelevance of science within everyday life. The popular concept of science as pure and disinterested thus makes it difficult to recognise within the realities of everyday life in these two communities.

At an immediately practical level, there seems to be a demand – not for a single authoritative source – but for a greater *plurality* of advice so that residents are not left with what they perceive as a monopoly of self-interested information. As one resident put it: 'Yes, they [the companies] are giving the right information, but I wouldn't know any different. Trust ... needs verification from outside' (Clayton resident). This suggestion that the industry can be trusted so long as this trust is 'verified' sums up well the local assessment of suspicion mixed with an awareness of dependence on that source – the company would not tell a direct lie (nor is it seen in an entirely negative light), but some institutional structure is needed to monitor their statements and activities at the formal, technical level. This point gains even greater significance if we now consider more carefully the motivation behind local citizens' requests for information and the kinds of response which they receive.

Knowledges in context; information sources and local needs

In our questionnaire, we asked that (relatively small) section of the sample that had actually sought out information on hazard issues 'What was it made you look for this type of information?'. Responses to this question seem to fall into a pattern if we categorise them across the three issues;

- on factory accidents/pollution
  - for health reasons
  - due to local accidents
  - noise
  - pollution/smell
  - nuisance
  - for study purposes
- on road traffic accidents
  - to stop people speeding
  - to stop traffic cutting down back lanes
  - to get action
  - to stop cars parking in the street
- on health and safety at work
  - concern about accidents
  - advice on how to tackle specific problems
  - as a requirement of the job
  - as part of a course

It seems very clear from the above headings (and also from the whole argument of this chapter) that 'information-seeking' fits into the social context of these two areas as portrayed here. The general picture was not one where requests were simply curiosity-driven, but instead where a different – and more 'practical' – pattern of motivations existed. Whilst curiosity may well play some part in information-seeking, a stronger factor seemed to be the demand for action or the registering of a complaint. Accordingly, a request for information was part of a larger process of attempting substantive improvement in a specific situation. Put differently, information-seeking emerges as an attempt to enhance one's control over everyday life and the external conditions which help shape this. In such a situation, the value of science may be somewhat less than advocates of an improved public understanding of science might wish.

This point can also be explored from the opposite direction – why do concerned people *not* seek out information? Whilst to the outside observer this may seem anomalous, apathetic, or indeed as a challenge to scientific evangelists to make 'science more fun', from the perspective of local citizens information will be of little relevance if it cannot directly tackle the underlying problem. 'I've complained about things all my life and it's brought me nothing but trouble' (Clayton resident). 'There's no point – you wouldn't get anything done about it.'

This well-developed social model of perceived powerlessness needs to be acknowledged as a basis for comprehending the everyday reality of two communities who generally view technical information as a means to an end (i.e. an improved living environment) rather than as an end in itself. Research in these two localities thus opens up for discussion the relationship between local people as *agents* and scientific understanding as a force for *change*. The general experience in these study areas seemed to be that technical resources served as an *obstacle* to social action (for example, in the direction of enhanced pollution control) rather than as a facilitator. In that sense, local people felt themselves to be struggling against technical information rather than actively employing it to help formulate and substantiate their demands and concerns. As Ulrich Beck has expressed this situation: 'A permanent experiment is being conducted, so to speak, in which people serving as laboratory animals in a self-help movement have to collect and report data on their own toxic symptoms *against* the experts sitting there with their deeply furrowed brows.'[4]

Certainly, the reassurances from industrial sources that 'everything was under control' contrast sharply with local people's detailed testimony of smoking chimneys, health complaints, noise, and smell. It is, therefore, important to consider in greater detail the *kinds* of information which are given to local citizens. What information response do local citizens consider themselves as receiving? Responses here with regard to factory accidents/pollution took the following main forms:

- safety measures being taken by the company
- assurances that the site is safe and that the problem is under control
- information on what the company does for the community
- nothing much/not enough

Typically, residents complained that requests for action/information were met with reassurances that 'everything is now OK' – thus leaving them no wiser as to whether there was a real problem at the site or not. 'The company says "don't hesitate to ring", they come round and apologise but they blame "problems" at the site – don't tell you what it actually is' (Eccles resident).

It would appear, therefore, that local sources of technical information (and, particularly, the company which is seen as *the* source of technical expertise) tend to couch any such information in a highly contextualised form so that it is more likely to offer reassurance or claims for the legitimacy of the source than to provide 'science' of the kind that working scientists would acknowledge as such. However, it is also important to recognise that there is no single 'science' to be tapped when it comes to issues of environment and safety – instead, particular 'sources' serve to provide particular views of uncertain and problematic expertises. Thus, for example, chemical companies may choose to play down uncertainties when addressing local audiences and to stress excellent working practices and recent safety records – often dismissing any recent releases as isolated 'problems' which have now been eradicated.

In that fashion, technical information as provided by industry to the community is constrained (or 'framed') within a broader social model based on the need to offer certainty and reassurance to an ill-informed local public. Within this model, 'science' serves as a broad legitimation for the information provided and for the authority of the source. However, technical analysis as such is not typically made available to the public – presumably on the grounds that they could not 'make sense' of this. Thus, technical analysis around risk and pollution is used as a general, but not a specific, rationale for the contextualised advice offered to local residents.

As we have seen, local citizens are likely to respond to this with the suspicion that they are being misled by a 'smokescreen of good will' (Clayton resident). More privately, industry officials will acknowledge the environmental problems posed by their location close to residential areas and express the view (with varying degrees of self-confidence) that – although they are operating to the best of their abilities – accidents can indeed happen. However, the industry view is generally that – at least in public – it is better to play down uncertainties than to feed local anxiety. Whilst different companies are likely to adopt slightly different postures in this regard – with one of the companies in our study area being notable for its attempts at 'open' community relations – there do seem to be social as well as technical constraints on the kinds of information which any single source can provide. These constraints can, for example, be a major obstacle when industry-agreed information about emergency procedures is distributed to the local population. Previous studies[5] suggest that industry's general message of reassurance and 'all is well' tend to distract from the need for citizens to react promptly to unforeseen incidents. Of course, this conclusion is particularly worrying when – as here – there is seen to be only *one* such technical source available.

Put at its strongest, therefore, local contexts such as those discussed in this chapter can produce a mismatch between the available 'sources' and the needs of citizens for reliable, trustworthy, and practically useful technical information which can then be supplemented to the available supply of local knowledge. In this case, we have seen something of the scepticism surrounding available sources of information. We have also seen the citizen dilemma whereby those sources which are perceived to be trustworthy may not possess the necessary technical competence, whilst those which do possess this are seen as self-interested (and hence untrustworthy). The pattern which then develops is that local residents see local industry as the only source of assistance, but feel that they receive only reassurances of safety (which often contradict local experience). Residents may feel 'short-changed' without possessing the technical competence to push industrial sources further. The consequence is that residents withdraw from information-seeking activities on the grounds that they are of little practical value compared to the local knowledges of everyday discourse.

It is important to note, therefore, that what may be seen as local ignorance (or local resistance) to technical information is an *actively* constructed social process rather than mere 'apathy' or 'irrationality' (a point which is further substantiated in Chapter Five by Michael). Local processes of 'making sense' of environmental threat may owe little to the statements of technical experts – but this is not simply a consequence of inadequate dissemination. If nothing else, therefore, this chapter has moved us firmly beyond the notion of a public 'deficit' or 'failure' with regard to scientific understanding. Instead we have seen much more dynamic processes of knowledge creation and assessment at work.

Such an analysis of the local treatment of environmental issues has, quite clearly, taken us a long way from the abstractions of the 'public understanding of science' as it is conventionally presented. In particular, our location of 'public understanding' questions within a specific social and cultural context has allowed the relationship between formalised science and everyday understanding to emerge in a manner which is altogether more problematic than generally portrayed.

In the final section of this chapter, we will turn briefly to a more critical analysis of technical information and its perceived relevance to contexts such as those described here. Once we have 'recontextualised' the public understanding of science, what place is left for scientific information and advice within everyday life? In discussing this point we shall, of course, be especially aware that this case-study deals with one highly loaded social situation and a particular example, rather than a necessarily typical – or even ideal-typical – case. Nevertheless, other chapters in this collection suggest similar themes and issues.

Discussion

In many ways, the analysis above can be seen as offering a gloomy picture to those who are keen to proselytise science and to advance its popular understanding. Whilst from an outsider's perspective, science and technical information are central to the problems faced by these groups of citizens, very few references to science or scientists were made during our interviews without specific prompting. Furthermore, questions about science tended to be met with either puzzlement ('how did we get on to *that* subject?') or fear and anxiety as dim classroom memories come to the fore ('I've never been very good at that kind of thing'). Thus, it seems that the closer one gets to everyday local discussion of technically related issues the more likely it is for science to 'disappear'. Put differently, the more one becomes aware of the realities of everyday life among particular communities, the more stark the disparity appears between the agenda and perspective of 'outsiders' (for example concerned scientists) and those of 'insiders' (i.e. those who actually live with specific worries and concerns).

In the early stages of our investigation, this phenomenon caused some frustration since we assumed that the low profile given to science represented a failure in methodology – so that science was somehow slipping through our empirical net. Later research – especially within semi-structured interviews – revealed, however, that science (whether as a social institution or as a body of expertise) was indeed seen as of limited relevance to the problems at hand. This point also suggests the limitation of general social surveys in this area – it seems unlikely that a broad questionnaire could capture the dynamics of community life.

Nevertheless, we would argue that notions of 'science' and technical expertise play an important role within local discussions of hazard. In particular, the superior technical expertise possessed by industrial groups permitted them to offer a message of certainty and reassurance. However, the 'science' presented by industry was seen by local residents as a very particular form of 'industrial expertise'. Whilst this could not be directly challenged by the residents, it was generally received with scepticism and caution. Thus, science does not simply parachute itself into specific social contexts – instead, it is constituted, reformulated, and framed according to the context of its dissemination (a theme which will be further explored in Chapter Three). In such a situation, public discussion was less about the 'facts' of pollution than the social and industrial processes which lay behind that pollution. 'Making sense' of hazard information thus meant fitting 'industrial expertise' into a pattern of preexisting knowledges and understandings.

Of course, this point about 'disappearing science' can also be made in the converse direction. Whilst from the perspective of scientific groups the problem may well be one of 'disappearance', from the perspective of citizens it is highly artificial to pick out one aspect of their everyday life and label it 'scientific'. Such categorisations lose significance within the larger social and cultural context portrayed above.

This analysis has also suggested the very critical reception which is likely to be given to any efforts at disseminating scientific information within a 'loaded' social context. Far from adopting a stance of 'all information is equally good', specific audiences will scrutinise the source very carefully. They will also assess the relevance and applicability of the information. In this context, of course, it must be remembered that a doctorate in Chemistry may be less useful than a knowledge of who to contact, how to organise and when to lobby. Where information and instrumentality (i.e. getting things done) are closely linked, science is not necessarily the most appropriate source of understanding and expertise. Similarly, where social powerlessness is perceived to be high, efforts at information dissemination may achieve low success rates unless they also offer the possibility of social empowerment.

Therefore, in contrast with much of the rhetoric surrounding the public understanding of science, our study suggests that science has no special status for everyday life, but must instead compete with all other sorts of knowledge and understanding (especially those categorised here as 'local' knowledges). Very often, the 'test' for the applicability of these knowledges is the extent to which they assist in the understanding and control of one's life. Accordingly, information which is seen as highly pertinent to certain social groups (for example, reassurances from industry that everything is under control) may appear irrelevant and diversionary to other groups and individuals. Going further, it is hardly surprising that science has a negative connotation for many citizens when those requests for information which are put forward tend to lead to a sense of dissatisfaction and disadvantage – and hence reinforce rather than overcome feelings of social powerlessness.

Put bluntly, science as encountered by these citizen groups is far from empowering or conducive to active citizenship. Instead, it serves to substantiate the social stance of certain relatively powerful groups – to consolidate the disempowerment of 'local' people. Whilst scientists may rightly object that this is not *necessarily* the case, the current relationship between science and social power cannot be ignored. Indeed, as this whole chapter has argued, this relationship is central to the public understanding of science – at least in this context of community pollution concerns.

In making these points about science in the local context, it is always necessary to stress the point that 'science' is not one homogeneous category, but is itself a dynamic and divergent body of understandings. Thus, Macdonald in Chapter Seven of this book discusses the presentation of scientific controversy in one context – a museum exhibition. Similar issues appear in this chapter as groups such as industry 'frame' science in order to offer a largely reassuring and 'authoritative' account of environmental threat rather than one which emphasises, for example, the inherent problems of juxtaposing hazardous industry and housing or the probabilistic nature of accidental releases. In our study areas, no alternative scientific account of the situation has been put forward – so that only one technical interpretation has effectively dominated. We have seen above that local residents are not naive to the implications of this partial hegemony – indeed, they operate with a relatively sophisticated model of science as representing social interests. Thus, for example, it is a routine local observation that technical information is presented in order to obscure the underlying causes of pollution. However, this awareness has not solved the practical problem of finding alternative sources of expertise in order to hear a greater diversity of scientific evaluations.

None of the above points should be construed as an argument against the communication of science to the general public or as a rejection of the relevance of science/technology to everyday life. Instead, our attempt here at 'recontextualisation' is intended – through the establishment of a more accurate and sociological account of citizens and sources than that usually offered by the 'public understanding of science' – to serve as a precursor to practical initiatives in this area. Whilst particular attention will be paid to such practical measures in the conclusions of this book, a number of brief points can be made here.

First, the need for more than one source of technical information within a given context has been stressed within this chapter – such alternative sources need to be accessible, local, and sympathetic to the needs of citizens. Secondly, implicit in this whole chapter there is the case that – if scientists are sincere in their desire to communicate more effectively with the rest of society – then this will involve a willingness to engage with alternative worldviews and 'knowledges' rather than labelling them in advance as emotive and ignorant. Certainly, we would argue that 'public understanding' needs to be viewed as an interactive rather than narrowly didactic process. Thirdly, as will be argued in the Conclusion, such changes in the 'social and intellectual humility' of scientists will also involve a reappraisal of their own institutional structures – a reappraisal which may well have an impact on the kinds of knowledges which they develop and disseminate as citizen voices begin to be heard within the science policy process. In that way also, serious commitment to the enhanced public understanding of science can only develop alongside an equivalent commitment to the enhanced scientific understanding of the public.

Finally, we need to consider the wider analytical applicability of this case-study to other contexts of science–public interaction. That is a theme which will be developed in the chapters that follow.

NOTES

1. The Royal Society, *The Public Understanding of Science* (London; The Royal Society, 1985) p. 12.
2. Williams, R., *Resources of Hope* (London and New York: Verso, 1989) p. 8.
3. 'Bricolage'; 'learning by doing', by imitation and improvisation, usually by oral as opposed to written communication. Levi-Strauss, C., *The Savage Mind* (London: Weidenfeld and Nicholson, 1966).
4. Beck, U., *Risk Society; towards a new modernity* (London, Newbury Park, New Delhi: Sage, 1992) p. 69. For a fuller discussion of the relationship between this case and the 'risk society' see Irwin, A. *Citizen Science* (London: Routledge, 1995).
5. Jupp, A. and Irwin, A., 'Emergency response and the provision of public information under CIMAH – a community case-study', *Disaster Management* 1, 4 (1989) 33–8.

3 Disembodied knowledge? Making sense of medical science

HELEN LAMBERT AND HILARY ROSE

The cloth of meaning may have to be woven out of a myriad scraps and off cuts, but woven it is, day after day, year after year.[1]

This chapter addresses the issue of public understandings of science through a study of the ways patients with a genetic metabolic disorder make sense of the medical sciences they encounter. This disorder is 'peculiarly scientific' in that although it is strongly associated with premature death from cardiac arrest, it does not in itself produce obvious, subjectively discerned symptoms. Indeed, as far as the patient, and most probably his or her general practitioner, are concerned, even those external indicators of the disorder that may be present are unlikely to be seen as significant. Thus prior to the onset of coronary heart disease and/or a heart attack, the disorder may have no embodied presence in the daily life of the individual; instead it is called into existence through laboratory indicators, that is by the presence in a blood sample of raised levels of lipids (blood fats).

As a science concerned with human health, medicine has played a crucial role in representing science and technology in general to the public and in establishing their authority and status in contemporary society.[2] While medical sociology has debated the sociological value of studying those whose disorder is medically defined, few studies have directly explored patients' interpretations of medical science itself. For social scientists interested in the public understanding of science, patients with such an abstract 'disembodied' disorder offer a rich point of inquiry. How, to return to our title, do people make sense of knowledge from medical science which does not look to, or speak about, the body?

The (disembodied) account from the sciences

The genetically transmitted inability to metabolise lipids – soluble blood fats which include the widely known cholesterol as well as the rather more obscure triglycerides – and its associated outcomes has been long recognised in medicine, but, because there has been until relatively recently no effective therapeutic response, there has been little clinical interest in diagnosing patients. This lack of interest signifies a tacit medical ethic that it is inappropriate to draw someone's attention to a health problem unless there can be an effective therapeutic response. (A similar stance among Cumbrian sheep-farmers, who saw no merit in post-Chernobyl whole body scans as there is no effective therapy to deal with high caesium levels, is described in Chapter One of this book.) Today however, a range of drug treatments together with the careful observance of a dietary regime offer the possibility to the clinician, with the patient's cooperation, of bringing lipid levels down to something close to, or within, the normal range. This past absence of diagnosis for a known disorder was reflected in those we interviewed, in that there was only one brother and sister with recollections of the pretreatment era, where a doctor parent had understood the nature of his and his children's condition.

There are several types of familial hyperlipidaemia that have been identified as 'inborn errors of metabolism' in medical research. Only one type, Familial Hypercholesterolaemia, is generally accepted as being due to a single gene defect that is associated solely with raised levels of blood cholesterol. Its molecular genetics and metabolic characteristics have been explored in a significant body of research for which the Nobel prize in physiology and medicine was awarded to Goldstein and Brown in 1985. This inherited condition, which is associated with premature death from heart attacks, constituted an obvious focus for research into the regulation of cholesterol metabolism in that its inheritance had already been thoroughly mapped in particular families, showing that it was likely to result from a dominantly inherited autosomal single gene defect. Although the common heterozygous form of the disorder, which produces blood cholesterol levels of roughly 50% above normal, is often without any visible symptoms, it may be indicated in adults by the presence of one or more of the following external signs: xanthomata – lumps of deposited and calcified cholesterol on the tendons of the back of the hand and the Achilles tendons; corneal arcus – a white ring round the cornea more normally seen in the eyes of elderly people; and xanthelasmata – yellow fatty deposits of cholesterol around the eyes. Only tendon xanthomas are a 'diagnostic hallmark' of Familial Hypercholesterolaemia (henceforth FHC); the other signs may be found in people who have raised blood cholesterol levels without having this particular inherited condition.[3]

In its heterozygous form the incidence of FHC has been estimated at 1 in 500 in the general UK population and it is therefore considered, despite its low public profile, to be one of the most common genetic disorders. Its incidence, as with other genetic disorders such as cystic fibrosis and thalassaemia, varies between groups of different geographical and ethnic origin. As an autosomal disorder, the condition can be inherited by either men or women, though with rather different implications for morbidity and mortality in each. Because menstruation seems to confer some degree of protection on women, the onset of health difficulties typically occurs for men in their forties and women in their fifties. For the extremely rare homozygotes (an estimated 1: 1 000 000) who inherit a defective gene from both parents, the implications are very grave, with fatal heart attacks usually in the teens or twenties. Our study did not include any of these patients.

In addition to the intensively researched single gene defect of FHC, there are various other genetically influenced conditions affecting lipid metabolism. These range from other related 'familial' disorders such as Familial Combined Hyperlipidaemia (which has an estimated incidence of 1 in 300 persons in the UK) and Familial Hypertriglyceridaemia, which are strongly inherited through as yet unknown genetic factors within particular kin groups, to subtle combinations of unknown genes that predispose individuals (known as 'polygenic' inheritance) to develop hyperlipidaemia in interaction with envionmental factors. The implications for morbidity and mortality of these less clearly defined traits can sometimes be as serious as those arising from heterozygous FHC, and they may be more resistant to research and treatment. On the other hand, some types of inherited metabolic disorder which fall into the group known to clinicians and researchers as 'familial dyslipidaemias' may even protect against coronary heart disease risk, or may have no effect on it but produce other health problems instead. Naturally, only those abnormalities of lipid metabolism which give rise to physical ailments or increase an individual's risk of premature death are of importance to the clinician.

The account offered here of the scientific understanding of genetically transmitted lipid disorders is necessarily simplified. Inevitably it gives an impression of a clear-cut set of conditions that in practice cannot so easily be distinguished. For example, the description of FHC as a single-gene disorder tends to imply uniformity and predictability, whereas in fact it is both genetically and clinically heterogeneous, since there are many different possible mutations of the particular gene which produces FHC. Moreover, a new dominantly inherited single-gene disorder which mimics FHC in its effects on blood cholesterol has recently been identified, so that the latter no longer constitutes a unique case.[4] Indeed, classifying an individual patient's particular type of lipid disorder is often difficult and treatment usually proceeds by assessing changes in the lipid profile over time in response to dietary and drug therapy, rather than by reference to any definitive diagnosis.

None the less the hyperlipidaemias, and particularly familial hyperlipidaemias with FHC as the research centrepiece, are located at the crossroads of a number of important developments in medical science. Among these are the disciplinary advances in biochemistry, physiology, nutrition, and pharmacology which have direct relevance for the clinical management of these disorders. Other developments are technological, such as the automatic instant screening equipment whose implications, as the machines enter high-street chemists and health-food shops, are only beginning to be felt. Yet other important developments are more contextual than clinical. The most clear-cut is the health policy context in which reducing death from coronary heart disease (henceforth CHD) in the British population has become a major goal. Health promotion, with its emphasis on healthy life style and healthy public policy (see below), reflects and supports cultural shifts which are happening anyway as people's understandings of how to keep well and how to take care of their families change under a sea of influences.

The other major development, at present still a considerable distance from actual clinical practice, is an increasing focus in biomedical research on the genetic origins of disease. This development implies a future potential to modify genes, and research into FHC even now raises the possibility of genetic manipulation. Success in cloning some variants of the faulty gene has been reported, and recent successful work with a laboratory-bred strain of rabbits that have an analogue of FHC suggests the possibility of human genetic therapy, presumably on a homozygous patient, at some time in the future. The significance of these preliminary moves in the context of this particular hereditary condition, parallel with the search for the amelioration of a number of other serious genetic disorders, has to be located against the background of the globally orchestrated, multibillion dollar Human Genome Project, in which biology enters Big Science.

But while the research emphasis on the genetic basis of disease is intensifying, gene therapy has no part in clinical medicine today. By contrast, for patients with any of the familial hyperlipidaemias (henceforth FH), nutritional science and pharmacology are the important knowledges here and now. Knowledge of nutrition offers the possibility of modifying the environment within which the genes are expressed and thus gives to patients themselves some measure of control. However, unlike the raised cholesterol levels resulting from high saturated fat diets that individuals without a particular genetic susceptibility to hypercholesterolaemia may have, most forms of FH are not controllable solely through dietary modification. Treatment to reduce cholesterol levels and thereby reduce the risk of premature CHD usually requires drugs as well as diet. The effectiveness of the new generation of enzyme-inhibiting drugs (which block the production of endogenous cholesterol) is sufficiently impressive for lipidologists to make jokes about being able to take one's pills and then go and eat cream cakes.

Even though there are many debates within nutrition concerning the health value and implications of different foods, ranging from therapeutic nihilists to committed healthy-eating advocates, nutrition is generally portrayed in health education material and in the media as a key (if shifting) element in the promotion of a healthy life style. For most of us, nutritional science and its scientific definitions of what constitutes healthy eating enter everyday understanding jostling and competing with older understandings of what is good for you. Fish, which in a half self-mocking folkloric story was 'good for the brain' becomes, or at least the oily kind, 'good for the heart' in the medical-science story. This new understanding becomes available through an intense deluge of information reported in the medical press and the popular media, and is directed particularly towards women as the health carers within families.[5] The stories from nutritional science and from epidemiology, while they form a critical part of an attempt to foster healthy life styles through national and international strategies of health education and the promotion of healthy food policy, have also become part of the marketing strategies of the food industry as firms seek to persuade the consumer that their particular brand of low fat food, polyunsaturated margarine, or oatbran will protect against premature death from a killer disease. The encouragement through public policy, cultural change, and marketing strategies of a scientifically defined healthy lifestyle both produces a friendlier knowledge context for people with this kind of inherited condition and facilitates maintaining the diet by the greater availability of suitable food products.

'Understanding science' as an active process

The study on which this chapter is based was undertaken as part of a research programme in the 'Public Understanding of Science'.[6] As various chapters in this book suggest, inherent in the emphasis on the public's *understanding* of science in documents such as the British Royal Society's report, is an implicit assumption of passivity in the process of receiving and assimilating scientific information.[7] By contrast, one of the strongest themes emerging from interviews with patients in our study is the active nature of this process when it comes to making sense of, and coping with, their raised blood lipid levels. We found that most people actively apply their own general knowledge, clinical observations, and knowledge of personal and family medical histories to make sense of new medical information, and try to utilise it effectively and appropriately in the risk reduction strategies that constitute management of this 'disembodied' disorder. This is not to say that this group of patients is always exceptionally compliant with medical advice they receive, but that by and large they actively seek to get hold of what they understand to be the key elements from a deluge of often contradictory information from a multiplicity of sources, laden with different levels of prestige and trustworthiness.

Some two-thirds of the fifty patients we interviewed were drawn from the membership of a support group set up in 1984 as the Familial Hypercholesterolaemia Association, which then expanded rapidly to become a charitable organisation concerned with all the inherited hyperlipidaemias, adopting the name the Family Heart Association. Like the Cumbrian sheep-farmers who took part in the study described in Chapter One of this volume, the members of the Association were seen as likely, through personal necessity, to be particularly motivated to 'understand' the relevant science and this was a primary reason for taking their attitudes and interpretations as a focus of study. To provide a baseline of 'ordinary' patients against which this self-selected membership could be compared, a group of similar patients was drawn from a local outpatient lipid clinic.

For most patients, a primary objective in understanding the medical science relating to their condition is in order to use it in reducing their health risk within the context of their everyday lives. As such, patients tend to be quite selective about the areas of medical science that they wish to acquire knowledge about. Ziman suggests that this kind of relationship to scientific knowledge is instrumental,[8] yet a dichotomy between the 'intrinsic' and the 'instrumental' does not really capture the active way that people weave different strands of information into the cloth of everyday living. Nor does it allow for the fact that, although the initial impetus was risk reduction and thereby instrumental, for some patients knowing about the science became intrinsically interesting. Thus, although from outside we can prioritise different contributory domains of scientific and medical knowledge according to their direct relevance for personal management of the condition, this is not how patients weave the knowledge together. Among both groups of interviewees, what is regarded as 'relevant', 'useful', or 'necessary' for the management of this condition in terms of scientific knowledge and information varies greatly between individuals, and so does their cloth of understanding. All these people could be described as operating at a 'good enough' level of knowledge for their own purposes; yet what is 'good enough' varies widely according to the particular perspective of the individual in a specific situation.

> It seems the people whose expertise I'm tapping are mainly in the family, it's delivered free [laughs], then what I have noticed is the fish pushed more which may well be just because my brother and sister are being pushed that by their doctors and dieticians or whatever . . .

> I got professional advice from the hospital and we also joined the Familial Hypercholesterolaemia Association, FHA, who send diet sheets out, and we got diet advice from wherever we could, we got things that had been given to my mother and passed on . . .

As suggested by the above excerpts, many of the patients we interviewed employed informal networks to acquire new information. In particular, where several members of a family or kin group have been diagnosed as having inherited the condition, a number of them invariably discuss their treatment, compare their current cholesterol levels and their understandings of the nature of their condition, and exchange new information. This active process of learning and disseminating information belies the model of the isolated individual patient passively understanding or misunderstanding the information s/he receives from professional sources, and modifies even the more accurate image of the patient remaining ignorant through not being given relevant information in the first place. Thus we would view patients as 'health workers' actively seeking to understand and make sense of the sciences they see as relevant (following Stacey (1988) who conceptualises patients not as passive beings cared for by others, but as health workers actively contributing to the recovery or management of their health).[9] The way in which they do so is clearly related to the particular contexts of their everyday lives, and this dimension will be emphasised here.

Reading the representations of CHD risk

We have mentioned that several scientific and medical specialisations contribute to medical scientific knowledge about familial hyperlipidaemia; some of these focus on different forms of scientific evidence and hold divergent implications for health care. The fact that current science in this general domain does not offer to the public a unitary discourse has implications for FH patients' interpretations of biomedical knowledge and advice. While chemical pathologists working in this field may describe familial hyperlipidaemias as 'inborn errors of metabolism' and medical geneticists would discuss them by reference to the hereditary defects of DNA which code for these metabolic abnormalities, a third representation which is at odds with both of these is more familiar to lay people. In the health policy field scientific understanding drawn from epidemiology and nutritional science, rather than medical genetics or biochemistry, is foregrounded and consequently hyperlipidaemia is represented as one risk factor for CHD that is caused primarily by poor – but modifiable – dietary and exercise habits and obesity. Current health-promotion campaigns, as well as the increasing emphasis on 'healthy eating' and its promotion by commercial interests, present CHD as an essentially lifestyle-related disease that may be prevented through individual behavioural change.

For most affected lay people, FH is understood in terms of its effect on physical health, rather than its metabolic characteristics or genetic determination; it is associated with CHD and with increased risk of dying prematurely from it. In the health education context and in people's everyday lives, FH is difficult to represent as separate from other causes of raised cholesterol and individual susceptibility to CHD. A multiplicity of sources now provide general information or advice that is relevant directly or indirectly to the management of high cholesterol levels and the prevention of CHD.

Debates about CHD prevention, high street cholesterol screening, healthy eating, and so on are prominent in the media, and new scientific evidence relating to these issues is widely disseminated in the popular press. Many of those we interviewed referred to Sunday newspapers and television as sources of information and to books such as the *Eight-week Cholesterol Cure*, which advocates self-treatment with massive doses of oatbran and niacin to lower cholesterol levels. Women's magazines such as 'Good Housekeeping', 'Here's Health', and 'Women's Own' debate the importance of adopting a low-fat diet, while TV advertisements promote Flora margarine or Commonsense Oat Bran Flakes as cholesterol-lowering. References to the unhealthiness and dangers of the traditional fried English breakfast are made on the TV soap Eastenders, while a newspaper in Yorkshire – an area of high mortality from CHD – reported the results an epidemiological study under the headline, 'Low-fat diets "raise cancer risk"'. While this mass of often contradictory information is aimed implicitly at those who do not have a strongly genetic predisposition for raised cholesterol levels, the information is also seen as potentially relevant by those who have, as exemplified by the two following quotes from patients:

Q: I wondered if you could tell me about how you went about learning about it?

A: Well anything I can, I mean if I see a heading in a newspaper or anything like that I don't just say oh they're on about cholesterol again or they're on about diets and healthy eating, I always read it, and then if there's any publication like this Aim to beat Cholesterol Cure, I read it, even on packets of things if they're advertising Get our Healthy Diet or anything like that, I'll get it just to see what, if it's a load of rubbish or if they've got something you know that, you can never tell until you actually get it, it might be informative, it might be a load of rubbish, so I do sort of read anything that I can possibly get hold of on that line.

A: ... it's instinctive that you will veer towards somebody who's saying oh I've got this new sort of diet plan, or there's a new book on the market written by some crazy American who thinks he knows it all, so you will buy it and you will read it, even at the end of the day if you then scoff at what he's said, you will still be drawn to it, rather like a moth to a flame, if there's anything about sort of lipids, cholesterol, in the paper you will read it, what your end result will be you don't know, but you can't avoid it, so I suppose yes you do lean towards not a sceptical view but you might then feel sceptical afterwards, I don't know.

At the same time, specific and accurate information about *hereditary* hyperlipidaemia is hard to obtain. While emphases may vary between health education programmes which are aimed at lowering CHD morbidity at the level of the whole population to, for example, advertisements by cereal manufacturers which invoke new social concerns with healthy living to sell products, all these messages to the general public promote alterations in personal life style as the key to health and long life. It is not a matter of general knowledge that high blood cholesterol levels may also be inherited and can be resistant to alteration by diet and, for people with FH, the currently predominant popular representation that susceptibility to CHD results from self-induced risky behaviour contradicts the scientific understanding of their own condition that they acquire from medical specialists.

This feature is noted explicitly by some patients. A middle-aged man related how:

> I have read that it is inherited, or that certain things can be inherited, but there's certainly far more emphasis on diet and how diet can affect it . . . I've never seen any sort of explanation as to you know various theories on what you can inherit as regards to this particular problem, but I've just seen that you can but it seems to mention it very briefly without going into a great deal of detail, they seem to spend more time describing you know how diet can affect it, so I think you've certainly got this in your mind that it's diet, that is the most important thing about it, you know what can affect it is your diet.

According to the accounts of some patients, doctors too find the discrepancy between popular representations of reasons for susceptibility to CHD and the condition of FH contradictory. For instance, another fifty-year old man commented:

> And the first doctor I saw was, again quite, not surprised but he was saying Here you seem to be very fit, in all ways, taking the age; and therefore it was almost a bit of a surprise that that, you're not obese and yet you've got a high [cholesterol] level.

Thus the media, and the range of public and private medical and paramedical practitioners with whom affected persons come into contact, supply patients with differing information about, and interpretations of, current scientific evidence.

Given that those we interviewed had particular reason to be interested in information about cholesterol and the risk of CHD, it is hardly surprising that they tended to be well aware of current scientific opinion, at least with respect to certain aspects of the relevant sciences. Once diagnosed, most adults who have FH seem to regard the medical science and technology that is applied to treat them as highly beneficial (indeed, life-saving), especially where relatives have died prematurely from CHD. However, over time they appear to become aware of the provisional nature of some scientific knowledges, especially with respect to the areas of therapy with which patients are most familiar. These tend to be nutritional science – because of its perceived importance for self-initiated therapy, its immediate relevance for coping with the disorder in daily life, and its widespread dissemination – and the technical details of blood cholesterol measurement, since lipid tests are an essential and invariant part of their treatment and the only means of assessing their own 'condition'. Changes in dietary advice over time were often mentioned; for example the case of olive-oil, which a few years ago was completely proscribed to those on cholesterol-lowering diets, but which is now regarded as positively beneficial, was cited as an example of historical shifts in therapeutic guidelines by a number of people we interviewed. Again, many patients noted how the 'cut-off' points for recommended levels for blood cholesterol had shifted over the years, or were cited differently by different doctors. Such understanding of the historically contingent and partial nature of scientific knowledge in specific instances seems to be developed even though patients are inclined generally to accept as fact apparently authoritative statements from the representatives of medical science.

Medical science and embodied knowledge: the meaning of an asymptomatic condition

The public representation of science is generally mediated through its application or promised application; in the case of medical science this is through clinical practice. Most lay people learn about aspects of medical science directly through contact with medical experts and, because the focus of medical science is human health, have their own experiences of ill health to contribute to the understanding of medicine and the clinical encounter. While subjective experience usually enters into people's interpretations of medical conditions, the 'risk status' of the individual with FH, however, can be ascertained only through the results of laboratory biochemical analysis of blood samples. Except for those who have already developed symptoms of CHD (whom we tried to avoid among our interviewees) observational or personal knowledge held by the patient is taken to be irrelevant in the clinical management of FH. Accordingly, we were interested in how people without the personal experience of illness render meaningful the abstract and impersonal information about their 'level of risk' which is produced and imparted by medical science. The concept of 'disembodied' knowledge does not refer only to the question of how people make sense of the medical science that is held to be crucial to their physical well-being, but which does not speak directly about, or look to, the body. It also denotes the individual embodiment of medical knowledge that must occur in the treatment of this kind of condition – since intervention to reduce blood cholesterol levels consists in action on the body through ingestion of appropriate forms of food and pharmaceuticals, exercise, and weight control – and which must be achieved by the patient despite a lack of perceptual, direct indications of health status or of the effects of such action.

Two particular themes in the interview material well demonstrate this process of negotiation between the 'disembodied' identification and treatment of the disorder, and patients' necessarily 'embodied' understandings of it: first, people's responses to the diagnosis of a potentially life-threatening disorder of which (among those who had not already developed symptoms) they had no subjective awareness; and second, their understanding and interpretation of the technical information – blood cholesterol measurement – that is used to diagnose and chart the condition. On the diagnosis of FH, one middle-aged man commented that:

> It's frightening, it is frightening, you're going along thinking you're okay and not
> realising you've got this . . . [and] it's just invisible, it's just invisible . . . it's like . . . like

tekin an insurance policy or summin. You don't really know if you're gonna use it or not.

Another said:

> . . . you know I mean it's one of those things, you don't feel any different if somebody says you've got a high cholesterol level, you don't *feel* any different.

It is, however, precisely this lack of subjective perception that necessitates reliance on technical assessment through the results of blood tests, which some people find difficult to accept. Questioning the authority of medical science in this sense may be related to age. A 17-year-old girl, when trying to characterise the condition, used the word 'ill' and then went on to correct herself:

> Well it's not really that word, cos you don't feel ill with it, you don't get side effects from it, you don't look ill from it or feel ill from it. It's just that it is there, and if you don't prevent it or help it it can end up quite serious.

A 13-year-old girl who had inherited FH from her father and had a very high cholesterol level explained:

> I just don't think about it, I try and forget about it, because as far as I'm concerned I'm not ill.

For many patients, blood cholesterol measurements are the only evidence that they are at risk and they constitute the only means of knowing whether risk-reducing strategies are having any effect. They have to accept that the figures which result from lipid testing of their blood are meaningfully related to their future health status, although these measurements are peculiarly detached from their personal experience. The language used by patients to refer to cholesterol measurements is significant in this respect. A wide range of terms were employed to describe these biochemical measurements, including 'level', 'number', 'figure', 'count', and 'reading'. The terms used to compare one measurement with another also varied, some describing it as being 'lowered' or 'raised', others as 'up' or 'down', and still others as 'high' or 'low'. The terminology used in talking about cholesterol test results is related to individual perceptions of their meaning, and members of the same kin group tend to adopt a common language. The more fully the person participates in the biomedical model of the disorder and the meaning of test results, the more likely they are to employ the technical terms of biomedical discourse. The use of other terms of reference may indicate a different view of the relationship between test results and the person's present and future health status.

Accepting the significance of a raised blood cholesterol measurement also implies some understanding of the statistical link between blood cholesterol level and risk of CHD in the overall population. Yet some patients come to internalise, as it were, cholesterol readings so that they are interpreted as direct statements about the individual's current physical condition rather than as indications of relative risk. Although minor variations in cholesterol measurements are likely to be the result of laboratory-produced inaccuracies and are unlikely to mean anything about the person's immediate health, these figures are sometimes regarded as absolute criteria of well-being. One 17-year-old questioned precisely this tendency among patients; his following, quite lengthy, statement attempts to express his awareness of the distinction between statistical risk and individual prognosis, and also comments upon the reliance on numerical measurements in the control of cholesterol:

> I can understand this sastistic [sic] about people dyin' every 3 minutes from heart attacks. Someone in my family, my gran, one of my uncles died of a heart attack . . . But even still . . . nothing was ever pointed out to me, or to my mum in that fact, that it was actually, this problem of cholesterol. No one's ever pointed out to me, and *proved* to me, that it is cholesterol levels which caused that . . . They can say it increases your *chance*, of a heart attack, but no one can say, er, because you don't go on a special diet, and you don't take these tablets, you will have a heart attack next x years, because that would be untrue. . . . the diet that I eat now, is more balanced away from cholesterol than just about anybody I know . . . but because . . . since a young kid I've been . . . brought up that this count's important, you know I get stereotyped into thinkin that this count is the be-all and end-all. So what you do is you actually in fact, keep this count down. Um, so it's not really the cholesterol I'm fightin against, it's the count. I think that's the same for a lot of people . Cos people don't understand about the cholesterol all they understand is these numbers. And they go for the numbers.

Adults also express doubts about the meaning of the figures with which they are presented, explicitly acknowledging that some knowledge of the overall range is necessary since individual cholesterol measurements can be interpreted only in relative terms. For example a middle-aged man under specialist treatment for his raised cholesterol level described how, years previously, he had started watching his diet after having had a cholesterol test at his doctor's but had not known what his level should have been:

> I didn't really know about the limits and if the doctor had said to me it's perfectly normal beween four and eight, and yours is only just over the first time, and then when it went down to five, if he said well you know that's fine because of, but they don't, they just quote figures, I mean they could say to me you know you've got 50 in your blood and it wouldn't mean anything to me, I might think that's okay, I don't know what the normal limit is, apart from eight was obviously a little bit higher than it should be, but nobody's said what it should be.

None the less, this man had turned to library books and other sources of information and, despite the statement quoted here, went on to show that he had since acquired a thorough understanding of the nature and extent of health risk in relation to cholesterol level, other contributory factors, and the development of CHD.

Implicit knowledge

The last quotation exemplifies a response which we found to be very common among the patients we interviewed; people often claimed not to know about medical science, but their scientific understanding would subsequently become clear through contrastive statements about the attitudes or comments of other lay people. Of course, knowledge claims are related to the audience being addressed; to the interviewer's questions, patients sometimes professed ignorance or disclaimed knowledge (see the discussion of 'ignorance' by Michael in Chapter Five). However, the same individuals often displayed a relatively detailed knowledge and understanding about their condition in the course of talking about, for example, communication with other lay people (relatives, friends) or their views about awareness of cholesterol as a risk factor for CHD among the general public. Interviewees were asked routinely how they would describe or explain the condition to other people and this proved to be a productive way of eliciting people's implicit knowledge. For example:

> Well I've had to do on quite a few occasions trying to explain exactly why I was doing it, they tend to know about cholesterol because that's a word that's used quite a lot nowadays, but nobody has heard of triglyceride, they're not aware of that word, they don't know what it means . . .

Here the interviewee's specialist knowledge about different types of lipids emerges through contrast with the level of knowledge among unaffected lay persons. Knowledge about nutritional science was similarly revealed in this manner; the woman in her sixties speaking below demonstrates her more detailed understanding of the differences between various forms of fats than that held by her friends and her knowledge that not all margarine is unsaturated:

> I start by saying no animal fat which is the easiest thing . . . you tend to give up when it comes to explaining the difference between poly- and mono–unsaturates and they tend to say This will be alright for you, it's got margarine in, [I ask] Is it sunflower margarine or soya, [they say] Oh no it's Stork, and you say Well it's not good enough, [they say] Well it's supposed to be better for you than butter . . .

In describing how she simplifies her explanation of the nature of the condition and the dietary restrictions it entails when talking to unaffected people, a woman librarian in her thirties similarly reveals her own, more detailed understanding:

> In simple terms I say having to keep off animal fats, which is a much simpler thing for people to grasp, cholesterol still has a sort of medical ring to it, but saying keep low on animal fat . . .

In the following excerpts two women in their mid sixties and late thirties respectively indicate their own knowledge of the science, but have different views of the level of popular knowledge than the librarian:

> Well usually if you say FH it doesn't mean anything, if you say familial
> hyperlipidaemia they look so shattered, with the length of it, I just say it is a blood
> condition that you know is passed on to different people . . . I just say that it is
> something that is passed which affects your cholesterol levels, well everybody knows
> about cholesterol now don't they . . .

> It's, I think it all really depends on who you're talking to, as to how they're going to
> understand what you're trying to say, I mean some people obviously if I'm too
> technical it would go over their heads and they wouldn't understand, they can only
> sympathise with you. But mainly I tend to tell them that I have a blood disorder,
> because a lot of people don't know the word lipid, and they don't understand what
> that is, and it means that I manufacture too much cholesterol and I have to keep this
> under control by drugs and diet . . .

As suggested in the last excerpt, most lay people are not versed in the scientific
language employed to discuss lipid disorders and patients may therefore have difficulty in discussing their condition except with others who are already familiar with
it. While there are some differences in views about what constitutes 'common knowledge' among the interviewees, these quotes also illustrate in various ways the problems that patients experience in attempting to describe a condition that can accurately
be portrayed only in technical terms. Although only a few examples are given here,
the tacit nature of these people's knowledge was a characteristic feature of all the
interviews.[10] Interviewees tended to talk about, and thereby reveal, their own knowledge primarily by contrast with the understandings of others.

Related to the implicit character of patients' knowledge is the fact that 'science'
itself was very rarely mentioned at all. The knowledge that patients hold about their
condition appeared to be construed as different from medical science itself, regardless
of how accurate a representation of scientific knowledge their own understanding
was. Science itself disappears in conversations with ordinary individuals about their
own lives and circumstances; it has already been shown that specific medical knowledge is often implicit and perhaps what lay people themselves know, they do not
regard as scientific. 'Science' appears to be regarded as the knowledge held by those
who work in scientific fields; thus, by definition, what lay people – and especially
the general public – know cannot be science.

Heredity and CHD: understandings of genetic inheritance

The idea that blood cholesterol levels could be raised due to hereditary factors was
new to most of the people with FH who were interviewed. This is not to suggest
that patients did not recognise a hereditary aspect to CHD at all. Despite the lack
of emphasis on inheritance in health education about this disease, most lay people
recognise CHD to be a form of ill health and cause of death that may 'run in

families'.[11] However, even among those who claimed to be aware that CHD in general could be inherited, the hereditary character of FH tended to become significant only retrospectively, after the individual's own diagnosis of high cholesterol or of CHD. Commonly, an event such as a heart attack in a close relative first alerted those we interviewed to the possibility of a problem, although even this rarely led individuals to investigate their own risk status.

Although the fact that high cholesterol levels themselves may be hereditary had been novel to all the people we interviewed at the time of their diagnosis, and appeared to contradict popular assumptions about the lifestyle-related causes of CHD, most of them were able to make sense of this information by placing it within the context of their own knowledge about inheritance. These people appear to view matters of inheritance as entailing complex interactions of genetic, psychobiological, and environmental aspects. Their understanding is an active one shaped by comparison with other known inherited and acquired conditions, and by personal experience of inheritance patterns and susceptibility to CHD among kin. Their comments about the implications of this hereditary disorder integrate it effectively into a view of human life that acknowledges individual variation, the multifactorial causation of ill health, and the normality of human imperfection. For instance, asked whether it had been a surprise to find out that his cholesterol level was not raised solely because of his diet, a middle-aged policeman said;

> I don't think you can exactly describe it as a surprise, but it's something that you didn't give a great deal of thought to. But I mean I accepted it, I didn't think oh I don't believe that, I accepted it, amongst the many things you can inherit like the colour of your hair and your eyes and your temperament, but I mean to me you can inherit your temperament, you know and all things like that, not only your looks, so I'm not too surprised at it . . . just like as I say you can inherit certain physical characteristics, I think you can also inherit mental and obviously you can inherit bodily function ways, can't you?

Asked whether she saw having a high cholesterol level for hereditary reasons as different somehow from its being due to diet, a secretary in her late fifties replied;

> I don't know on the hereditary side of it, because of my hereditary [sic] I've got red hair, that's just one of those things, I mean there's nothing I can do about it, that's the way I was born . . . I don't feel the victim of circumstances particularly, I mean a lot of the things that I've inherited have been good things presumably.

A young woman with FH engaged to be married said that she and her fiancé had considered the implications of having children:

> Well we did think about that but, I mean if it can be helped it's all right. I mean it's not like um . . . the other things that can be hereditary . . . that you have to have tests for, like Down's syndrome or the other things like that . . . The only other thing that runs, in my family is twins. So not really because, it's an everyday thing . . . if you

have a child an the child's got high cholesterol, they'll eat a healthy diet . . . probably the only age you'll learn to eat the right foods an that, you know . . . It'd be like that.

Discussing how she described the condition to her 8-year-old son who has inherited FH from his father, a woman said:

> There seem to be more and more of these things, I suppose as genetic research goes on . . . I was explaining it to [her son] in these terms, it was almost the norm to have that sort of defect . . . because the perfect genetic set probably doesn't exist.

Scientific information about the genetic transmission of the condition is also, of course, understood in terms of the health consequences of the condition rather than in abstract terms. Thus, several people stated that boys were more likely to inherit FH than girls and quoted their doctors as the source of this view. Rather than construe this simply as a failure to understand the nature of autosomal inheritance, it seems likely (given other interview evidence) that, from the patient's viewpoint, the gene defect which is inherited is inseparable from its clinical consequences. Certainly with respect to both life expectancy and relative risk, females with FH are less susceptible than males at least until after the menopause. Clinicians are therefore likely to emphasise the importance of diagnosis and treatment in male rather than in female children.

Conclusion: situated knowledges

People clearly develop situated understandings of medical science through intensive experience of a specific domain (in this case, familial hyperlipidaemia). While these understandings may differ in significant respects from the received scientific/medical orthodoxy, such situated understanding adds meaning to formal knowledge by locating it within the individual's personal and social context. It is, moreover, continuously modified over time as new technical or scientific information is acquired and as personal and family health-related circumstances change.[12] This interpretive relationship to scientific knowledge is probably more valuable to the individual patient, and has closer parallels with scientific knowledge in its provisional contingent quality, than would a passive acceptance of medical science as univocal and absolutely authoritative.

The need to adjust to the disorder in a manner that is convenient and manageable in everyday life is, of course, the ultimate priority for those diagnosed as having FH. Thus the patients we interviewed accept the diagnosis they are given and follow medical advice, but most of them do so only to the extent that it can be accommodated relatively easily in their daily lives. People who had recently been diagnosed tended to express considerable anxiety about having FH, but those who had learned of the condition years previously no longer regarded it as threatening. Rather, as we have shown, it was seen as an aspect of their physical make-up which had required some adjustments in their lives but which, in the long term, they had found themselves able to accommodate. The 'scientific fact' of above-average personal risk is consciously weighed against its implications for a restrictive existence, and a compromise between risk-reducing action (through dietary control and so forth) and maintaining an enjoyable social life (in the broadest sense) is usually reached. People are not ignorant of, nor do they fail to understand, the technical assessment of their susceptibility; rather, they understand this and evaluate its implications for action with regard to the quality of their whole lives.

Similarly, among those interviewed we observed no clear correlation between educational level and adherence to medical advice. There appears to be no simple, causal relationship between the literal 'understanding' of scientific information, and the inclination or ability to act on it. Those who were most highly educated or at ease with scientific or statistical concepts were often most sceptical of medical interventions. They tended to question the assertions of medical science or relativise their personal risk against potential environmental health risks such as air pollution and traffic accidents, thereby reducing its overall significance.

Just as patients' personal understandings may be characterised as forms of knowledge that are shaped by, and situated in relation to, those aspects of their lives which are most affected by having a particular medically defined condition, so the various forms of scientific understanding relevant to the condition can also be seen as forms of situated knowledge. Rather than assuming science to be an objective, universal, and decontextualised knowledge that is refracted and particularised by lay people in their attempts to understand it, we have attempted to show here that scientific knowledge too is contested, provisional, and situated, with different understandings foregrounded according to the particular context in which it is invoked. The view of science's relation to the public as comprising decontextualised knowledge about which the public is deficiently cognisant[13] is challenged by the interpretive approach employed by this particular section of the public.

Yet the relationship between personal, context-dependent knowledges, and authoritative or institutional forms of scientific knowledge, is complex; the character of the former may result in enhanced trust in the latter or, on the other hand, in increased scepticism. In the case of medical science, such trust or scepticism finds expression in the extent and nature of compliance with medical advice. In turn, variations in the degree to which patients choose to comply with such advice provide some indication of how this particular public negotiates relations with a rather esoteric set of knowledges. It gives some encouragement to the view that the public – at least where they have reason to believe that knowledge can make a difference[14] and where there are multiple sources of information and advice some of which are seen as reliable – can be rather skilled at making sense of science.

ACKNOWLEDGEMENTS

We wish to thank the Economic and Social Research Council for funding the research on which this paper is based (grant number A418254002). The editors also deserve thanks for their encouragement and perseverance. Professor David Galton kindly checked the accuracy of the scientific content. We are grateful to the Family Heart Association and to Dr Andrew Grant for their assistance and co-operation and, especially, to the patients and families who volunteered their time and their insights with great generosity.

NOTES

1. Worpole, K. 'Total culture', *Marxism Today* (September 1990) 44–5, on 44.
2. Cornwall, J., *Hard-Earned Lives: accounts of health and illness from east London* (London, New York: Tavistock, 1984) p. 121.
3. Durrington, P. N. and Miller J. P., 'Clinical Aspects of Hyperlipidaemia', *British Journal of Hospital Medicine* (July 1984): 28–34, on 29.
4. Tybjaerg-Hansen, A., Gallagher, J., Vincent, J., Houlston, R., Talmud, P., Dunning, A. M., Seed, M., Hamsten, A., Humphries, S. E., and Myant, N. B., 'Familial defective apolipoprotein B-100: detection in the United Kingdom and Scandinavia, and clinical characteristics of ten cases', *Atherosclerosis* 80, 3 (1990) 235–42.
5. cf. Graham, H., *Women, Health and the Family* (Brighton: Wheatsheaf, 1984).
6. Our research was conducted between October 1988 and October 1990 in parts of England, under the auspices of the Economic and Social Research Council's Science Policy Support Group. For accounts of the overall research programme see Birke, L., 'Selling science to the public', *New Scientist* (18 August 1990) 40–4; and Ziman, J., Silverstone, R., and Wynne B., 'Special Report: Public Attitudes', *Science, Technology & Human Values* 16, (1991) 99–121.
7. Royal Society, *The Public Understanding of Science* (London: The Royal Society, 1985).
8. Ziman, J., *An Introduction to Science Studies* (Cambridge University Press, 1984) pp. 184–5.
9. Stacey, M., *The Sociology of Health and Healing* (London: Unwin Hyman, 1988) pp. 5–7, 194–205.
10. Cf. Ravetz, J. R., *Scientific Knowledge and its Social Problems* (Oxford: Clarendon Press, 1971) p. 239, who contrasts scientific knowledge as explicit and public, with personal understanding as private and largely tacit.
11. Davison, C., Frankel, S. and Davey-Smith, G., 'Inheriting heart trouble: the relevance of common-sense ideas to preventive measures', *Health Education Research* 4, 3 (1989) 329–40.
12. Ravetz (*Scientific Knowledge and its Social Problems*, 1971) suggests that:
    Each person's understanding of a piece of scientific knowledge, or even a fact, will be peculiar to himself, depending on his own history of involvement with the materials ... One can 'know' a classic piece of scientific knowledge from an early age, but one's understanding of it can, and should, develop as one matures. (pp. 239–40)
13. Bodmer, W., *The Public Understanding of Science* (The seventeenth Bernal Lecture) (London: Birkbeck College, 1986); Royal Society, *The Public Understanding of Science* (1985).

14. Anna Wynne's account of multiple sclerosis – where there are distressing symptoms but little science, rather like a mirror image of FH – makes some interesting observations about faith. Only the scientist with MS continues to have faith in the potential of science to cure MS; the others are more equivocal, not because of their lack of faith in the power of science but because they think it may fail in the specific case of MS. Wynne, A., 'Accounting for accounts of Multiple Sclerosis', *Knowledge and Reflexivity: new frontiers in the sociology of knowledge*, edited by S. Woolgar (London: Sage, 1988) pp. 101–22, on pp. 117–18.

4 Now you see it, now you don't: mediating science and managing uncertainty in reproductive medicine

FRANCES PRICE

Visual imagery is often presented as capturing 'literal reality', especially in a climate of increasing surveillance. The quest to better 'the view' has spurred the development of technologies that enhance visibility and provide useful images. Telescopes, microscopes, X-rays, and then ultrasound and other scanning devices have opened up wholly new areas of observation. They have also fostered new alliances, evaluation criteria and agendas – new routes to authority and of entry into the politics of representational practices. Surveillance, for instance, particularly since Foucault (1979), has become a way of asserting the political character of visualisation. However, this analysis may deflect attention away from imaging practices premised on a 'mutuality of knowing', and on 'care' as well as 'control' (Lyon 1993). Moreover, central issues of power and accountability also arise when there is a shift of domain, as the image is transferred beyond the legitimate realm of the 'expert'.

Images invite action: the use of technologies of visual depiction are now central to medical interventions. In general, diagnostic images are intended, above all, for expert scrutiny: elaborate conventions exist to shield such images from public display. Captured on plate, film, or tape, these images are scanned for signs of pathology and become substitutes, surrogates for the subject under clinical observation: as authorised depictions, they are also readily available for evaluation, in group discussion in the clinic, or in a court of law in a malpractice suit. Patients can be elsewhere: their physical presence is no longer required. The images become constituted as technical data, and such technical data can come to be seen as determining the diagnosis.

Yet, in reproductive medicine, women as patients are invited to view images of embryos and fetuses. The production of these images, which women view 'with their own eyes', is celebrated by clinicians as both an aid to communication and as encouraging their compliance. To this end, such images have come to be routinely produced in antenatal and infertility clinics.

Thus clinicians in reproductive medicine use imaging technology not only to enhance their diagnostic powers, but also to structure discussions with their patients. Clinicians 'help' women as patients to see, to make sense of the images before their eyes – they are accustomed to being authoritative as to what and when information is communicated, as part of the professional management of uncertainty. They create a frame, an enclosure for a narrative within medical discourse. Also, various institutional 'devices' are in place in the professional practice of medicine which enable clinicians to legitimately exert a focusing effect on behalf of their patients. In effect, they 'manage' the production of the image, its interpretation, and any information conveyed to the patient. But this institutionalised authority to set the agenda and to safeguard 'clinical autonomy' can compromise wider sociopolitical concerns, including the public understanding of science. In effect, medical practice can vest tentative scientific projections with robust authority.

Images produced for view by, and discussion with, pregnant women, and with those anticipating pregnancy, serve to shape knowledge and communication of information about the pregnancy, or about the prospect of a pregnancy, in ways that may ultimately prove misleading. The clinician's 'framing' may obscure the uncertainty of the science underlying medical practice. Visual displays – the practice of 'rendering visible' – can appear problematic in reproductive medicine in relation to the principle of informed consent, during both pregnancy and the quest for pregnancy.

The diagnostic 'rendering practices' – in other words, the institutional authority structures and the claims to autonomy of medical professionals – mediate the scientific practices involved. Ulrich Beck has commented on the successful professionalisation of medicine, such that 'medicine as a professional power has secured and expanded for itself a fundamental advantage against political and public attempts at consultation and intervention. In its field of practice, clinical diagnosis, and therapy, it not only controls the innovative power of science, but is at the same time its own parliament and its own government in matters of "medical progress"' (Beck 1992). The approach and scope of medical decision-making becomes of even greater concern as entirely new situations are created as reproductive and genetic technologies develop, opening up new opportunities for intervention. This has particular political repercussions for women because of the complexity of their incorporation into the polity as the 'different' sex, as 'women', as those with the capacity of being 'mothers', in contrast to men the 'individuals', the 'citizens' of political theory (Pateman 1992).

The focus of this chapter is the production and interpretation of images of embryos and fetuses in the context of communication about them. I draw on recent fieldwork: a project that was part of the National Study of Triplets and Higher Order Births (Price 1989, Botting et al. 1990, Price 1992)[1] and a supplementary project to the National Study. Both projects involved interviews with clinicians and scientists and with women and men attending infertility clinics (Price 1993).[2]

Rendering visible

In the Western world, vision has been privileged over other senses: the practice of image making as a route to scientific discovery has a long history (Arber 1954; Rorty 1979; Keller and Crontkowski 1983; Lynch and Woolgar 1990). 'Seeing' in the sciences is a distinct activity. Consensual ways of 'seeing' and 'knowing' are maintained socially through shared paradigms.

Scientific communication typically involves the use of visual displays to simplify and schematise objects of studies: such 'rendering' practices are necessary for constituting manageable data (Lynch 1990).[3] By producing technical data from visual displays, scientists can construct and compare models of what can be seen. To achieve this, they compare images with images, and translate information acquired from other sources into the images obtained by a new device (Pasveer 1990).[4] This all takes time. When observations are novel, scientists have to decide what should be regarded as important and what as artifact (Lynch 1985).[5] It is this very indeterminacy of the data which provides room for controversy in scientific practice.

The medical historian Stanley Reiser has traced the development of the tendency within the medical profession to regard illness in terms of 'discrete, picturable lesions, as a disturbance of one part of the body more than of the person himself'. He also charted the growth of the idea that 'the most significant advances in diagnosis would come from new ways of visualising pathology' (Reiser 1978: 55). When the doctor could see and make sense of what was within the body, it became less urgent to listen and make sense of what the patient reported. The voice of the patient became muffled, and the task of medicine became not the elucidation of what the patient said but what the doctor saw in the depths of the body (Armstrong 1984: 738).

Reiser links the twentieth-century quest for machine-produced evidence in medicine to a faith in science and technology, and 'a belief that a scientific spirit entered clinical practice through technology' (Reiser 1978: 161). Once they came to be largely reliant on state-of-the-art technology to diagnose and monitor illnesses, doctors could align themselves (by association) with a model of science incorporated in contemporary technical expertise, in which science is seen as generating empirical truths through rigorous methods. In so far as images can be 'objectified' and used diagnostically, the clinician's authority over the patient, and also other professionals, is bolstered: the intended outcome is predictability and control. The focus is the image itself.

Clinicians constantly use images as visual clues to devise appropriate ways of treating their patients – so much so that a leading American medical journal recently asked its readers to send interesting photographs of their patients' ailments. 'As doctors we encounter an enormous variety of images from day to day', wrote the editor-in-chief of *The New England Journal of Medicine*, Jerome Kassirer, introducing a new feature, 'Images in Clinical Medicine' (Kassirer 1992). He welcomed the submission of 'typical images', in colour or black and white, of skin lesions, funduscopic views, blood smears, ultrasonic scans and more. Yet he acknowledged that 'an image considered typical by one expert may seem atypical to another'. Such conflict would stem in part from 'differences in judgment, training and experience'; disagreements will be adjudicated by peer review. This will help to ensure, he argues, that the images selected are 'prototypes', to be usefully 'anchored' in the readers' memory as 'diagnostically important objects'.

Such a focus, however, risks losing sight of not only the practice that has created it, but that this practice takes place within a particular social organisation of technical skills. Image-making in medicine is a collaborative venture with others whose expertise comes from training in scientific and paramedical occupations including 'third party' professionals with appropriate skills. Yet the status of all these collaborators is subordinate to that of the clinician (Daly and Willis 1989). In the UK, the occupational ideology of medicine legitimates a hierarchy of expertise.

Thus in contemporary practice, the operator of the device produces images – Michael Lynch's 'docile object' or 'material template' – which can then be worked on by 'experts' (Lynch 1985). As sources of information, prediction and planning, the displays from imaging devices thus become workplaces and sites of discovery, contention, and negotiation – the familiar 'rendering practices' of scientific inquiry. But the context of application is clinical practice: in important respects, this is not science. Moreover, the technical division of labour between the execution of the imaging task and clinical practice, combined with authority and status distinctions, ensures overall medical dominance (Willis 1983). Effectively there is a hierarchy of credibility which is of crucial significance in the management of uncertainty.

Judgements about whether to proceed with a treatment or diagnostic examination, are, however, the responsibility of the clinician, as are decisions about how to account to patients about the attendant degree of uncertainty and risk (O'Brien 1986).[6] Doctors decide what risks their patients can bear. In English law, patients must be told of the *substantial risks* involved in medical treatment (Sidaway v Gov. of Bethlem Royal Hospital, 1986). But how much a clinician should disclose is left as a matter of clinical judgement. Clinicians are assumed to employ their medical expertise as the agents of their patients in the attempt to maximise the benefits of treatment.

They are presumed to be aware of the range of patients' preferences – one such preference being that of minimising risk. But risk is a culturally conditioned idea, shaped by social pressures and notions of accountability (Douglas 1992). Clinicians' priorities may not be those of their patients.

The presumption that clinicians can know what risks their patients can bear takes on a heightened sensitivity in relation to infertile women and men who, as Naomi Pfeffer and others have emphasised, are commonly represented as 'desperate' and willing to go to any lengths to have a baby (Pfeffer 1987, Pfeffer and Quick 1988). 'The patient' is, in most cases, a couple and both clinician and 'the patient' perceive pregnancy to be the goal. This focus on pregnancy at all costs is exacerbated by the contemporary customer service model of health care provision. Also the provision of infertility treatment is largely in the private sector of medicine, and clinics compete in the marketplace for customers on the basis of their published pregnancy rate.

Looking at embryos

The quest for diagnostically useful images in reproductive medicine has encouraged novel ways of viewing many stages of human development including ovulation. Since the laparoscopic collection of oocytes for *in vitro* fertilisation (IVF) and subsequent embryo transfer, the very earliest stages are open to scrutiny (Betteridge 1981, Westmore 1984). Several pioneering partnerships of clinician and scientist developed the IVF technique (Edwards et al. 1980, Yoxen 1988). Subsequently IVF teams have been set up around the world, forging close working relationships between laboratory and clinic which then became sites of collective action.

The practice of IVF is critically influenced by the uncertain scientific status of 'quality' assessments based on the visual appearance of embryos. The major factor affecting the pregnancy rate in the IVF procedure is widely held to be embryo 'quality'. But the factors implicated in embryo implantation in the uterus are still speculative (Seppala 1991; Turner et al. 1994).[7] Current assessments rest on examining magnified images of embryos and 'grading' them according to 'visual estimates of fragmentation and the evenness of blastomeres', according to the embryologists Jennifer Hartshorne and the IVF pioneer Robert Edwards, both working at Bourn Hall at the time of writing. Embryos with misshapen or damaged cells are suspected of being less likely to implant in a woman's uterus, but it is acknowledged that appearance has little predictive value: most forms of embryonic morphology during early cell division have scant relationship to subsequent developmental capacity.

Researchers in various laboratories in Britain and abroad are attempting to find biochemical indicators of 'embryo quality', but as yet no reliable marker has been discovered (Leese 1981, 1989; Leese et al. 1990; Martin et al. 1990; Turner et al. 1994). So the visual assessment of embryo morphology continues, although the results of placing different 'grades' of embryos in the uterus turn out not to be greatly different 'unless substantial fragmentation has occurred' (Hartshorne and Edwards 1991).

This grading process rests on years of 'looking at embryos': the embryologist's practice appears more as an art than a science. The embryologist Karen Dawson explained (in a BBC Horizon television programme) the grading system she used:

> OK, so here are three embryos that we're going to transfer today. The one at the top is a four-cell embryo, one of the cells has started to fragment and it would be a grade 2.5 on our grading system. The one in the middle is very nice. There are 4 cells, they're all still intact, and that would be a 1.5 on our grading system, and the bottom embryo looks as though there was a fifth cell which has started to fragment; again, it's quite a nice embryo although the cell fragmentation does take it down to a grade 2. All you can use is your experience and your knowledge at looking at embryos and sometimes you can't really explain to somebody why you particularly pick one embryo in preference to another; it's just a feel that you get over the years of looking at embryos.[8]

Ultimately, the criteria for embryo selection are arbitrary unless the embryo is virtually disintegrating. No one can be sure that a viable embryo has been chosen. So IVF teams routinely transfer more than one to a woman's uterus in an attempt to achieve a 'good-enough' pregnancy rate. When three or more eggs or embryos are available for transfer, three are generally transferred; this has come to be regarded as standard practice.[9] But this practice is controversial because of the known risk of multiple pregnancy.[10] Neonatal paediatricians in particular have deplored the incidence, risk and consequences of multiple births and their effects on neonatal services (Levene 1986, 1991; Anderson 1987; Peters et al. 1991; Scott-Jupp et al. 1991).[11]

In the face of this uncertainty embryologists have advanced stories that have appealed to clinicians as practitioners. One was the idea that there is a synergistic effect when more than one embryo is transferred – which Edwards, employing a surely misplaced 'caring' analogy, called embryo 'helping'; and Edwards has long since retracted this conjecture (Edwards 1985). Such comfortable explanations, still reiterated from time to time, enabled a consensus to be reached about what 'good-enough' practice should be in the circumstances.

What started as a pragmatic decision to agree a number then becomes difficult to shift. In this way, state-of-the-art embryology applied in the clinic becomes locked up in medical practice. Transferring three embryos has become the norm (World Collaborative Report 1993), despite the fact that several embryologists have published data that suggest that transferring only two would reduce the risk of a multiple pregnancy without lowering the pregnancy rate (Bennett et al. 1989; Dawson et al. 1991; Waterstone et al. 1991; Staessen et al. 1993).

The staff of one IVF clinic set up a randomised clinical trial in the hopes of producing a scientifically rigorous demonstration that would confirm their belief that the pregnancy rate following the transfer of two embryos was as good as after the transfer of three. But for this very reason the clinic found it difficult to recruit women for the trial who were prepared to accept the transfer of three embryos:

> We're actually finding it difficult to find people to go on our trial because so many of them want two [that is, they do not want three]. Um, the problem is, you see, we do tend to say to them that we feel that two gives as good a result as three. Because if you want to put people onto a trial you've almost got to say that, otherwise you, if they thought they'd get a lesser chance, they'd go for three. And you'd never get anyone on the two side. So it's quite hard to pitch your counselling right, I find.

Perceptions of acceptable risk

In the IVF procedure, unusually in medical practice as a whole, clinicians generally favour allowing patients 'to see with their own eyes' what is going on. Images of embryos, graded behind the scenes by the embryologist, are typically presented to the woman, with their grading, as part of the procedure: 'These are the embryos we're going to transfer today.' Moreover, a woman may find the sight of her embryos compelling:

> Three grade 1 four-cell embryos were transferred to my womb. I lay on the table looking at the magnified embryos on the television screen above my head. The four-cells in each of the embryos were clearly visible, and it suddenly struck me that I was looking at three potential human beings, a combination of [my partner] and myself, and they seemed very beautiful. Tears started rolling down my cheeks and the nurse gave me a tissue and rested a reassuring hand on my shoulder.[12]

This woman vividly anticipates what each embryo might become. Yet she is 'seeing' the embryos as 'out there' as detached from her, as part of the plan for pregnancy. She views them, magnified on a screen, as individual, free-floating entities. It then becomes difficult to change perspective.

One clinic, at least – at the Hammersmith Hospital in London – provides photographs of these embryos for women to take home with them. The senior counsellor Jennifer Hunt explained that women welcomed the photographs, and that it helped them if no pregnancy resulted to have evidence that they had 'got to the embryo stage'.

People come to IVF clinics with expectations about what will happen there, and often do not wish those expectations to be disturbed. Deferring to the judgements embodied in the current practices of the IVF team, the possibility of triplets may be regarded as an 'acceptable risk' of an assisted conception procedure (Price 1991). A man attending one of the private clinics with his partner who was awaiting IVF, acknowledged that this was his position: 'I would imagine that a lot of people would be in our sort of position. They'll take it [the possibility of triplets] as an accepted risk, without really giving it much thought.' He made it clear that he felt ambivalent about the idea of any information which jeopardised this sense of acceptance: 'We're paying for this treatment and we're paying to come to experts and if experts think that that's [the transfer of three embryos] the best thing to do, well, that's what we'll do.'

He was satisfied with the level of communication about the procedure. Such data as there are suggest that it can be inordinately difficult to switch context, to envisage consequences of an IVF procedure projected beyond the confines of the clinic – in effect, one context 'decontextualises' another.[13]

Most of the women and men attending infertility clinics whom I interviewed welcomed the prospect of having twins: they had absorbed the idea that twins are a possibility after IVF from newspaper accounts, friends and from the clinics themselves. Photographs of twins born to women after undergoing IVF line the walls of rooms and corridors in many clinics. That there were risks attached to this outcome came as a shock to some of those interviewed (Price 1991). One woman described her surprise on being told by clinic staff of the risks of a twin pregnancy:

> We knew that there was a risk [of a multiple pregnancy] because of the press and things making you aware of it. But we were actually – well, I was really thinking it would be a wonderful thing to have twins. And *then* they said at the [clinic's waiting list] seminar that they don't like multiple pregnancies because it's dangerous! So I was thinking about it more.

Other women and men attending infertility clinics adopted broader 'acceptance' strategies. They denied the need to make enquiries beyond what they were told by the clinic. They, like the apprentices at Sellafield, interviewed by Brian Wynne and his co-researchers about their knowledge of radioactivity, did not want to know more (Wynne 1991), or 'to go into it', as one man whom I interviewed put it:

> Some people just seem to go along [with it] like we do. We just sort of go along. We don't – we get told everytime we come to a meeting, you know, what's going to be happening next. *We don't tend to sort of go into it. (Emphasis added).*

His partner added: 'We think it's almost superstition in a way. I don't want to talk about it, because at the end of it, if nothing happens'. This strategy makes these couples particularly vulnerable if there is an adverse outcome. Those people in the National Study who reported that they had accepted the IVF procedure without question were the least prepared for triplets or quadruplets because they had never envisaged the possibility. They may be greatly shocked by any untoward consequence of treatment.

Their vulnerablity stems in part from the fact that they are tacitly encouraged to trust that untoward outcomes can be anticipated and remedied in clinical practice. This vulnerability is further compounded by the extent to which they come to perceive the procedure (including the decision about the number of embryos to transfer) as the responsibility of the clinician – and so contextualised within the medical frame.

One example of such apparent detachment from the possible future familial implications of the IVF procedure comes from a woman in the National Study who was told at her first scan that all four embryos that had been transferred had implanted. She spoke of the experience as 'bizarre'. 'You are not able to analyze your own feelings' she said 'It's something *so weird*'. Her partner added: 'For quite some time we didn't know what that meant. Whether it meant that eventually there would be four. Or whether it meant that they would, as it were, be discarded along the way?'

Viewing their embryos seemed to encourage patients to see the transfer of three (or in earlier years, four) embryos as simply part of the medical procedure; the link between the entities in view, and the prospect of a high-risk multiple birth, and its relational consequences, are obscured. The focus for both clinician and patient becomes the facilitation of a pregnancy, and the procedure of transferring more than one embryo comes to be regarded as an acceptable risk.

Ultrasound images of the fetus

Fetal images, and information disclosed about them, may be of profound significance in the lived experience of pregnancy (Stewart 1986). However, scans are undertaken, first and foremost, to inform the clinical management of the pregnancy. The focus on obtaining a good-enough diagnostic image for 'appropriate patient care' lies at the heart of the use of ultrasonography in obstetrics. In this context, what is visualised through this 'window on the womb' is controversially regarded as a patient within a patient (Strong 1987). The second patient is modelled in sound waves on a monitor: the screen becomes the workplace for diagnosis and, sometimes, therapy. Through this medium, the fetus comes to be regarded as 'knowable' by the clinician, in a way that is seen as detached from or independent of the pregnant woman (Petchesky 1987). New 'responsibilities' arise as the fetus is constructed as having a claim to separate medical attention, which in turn alters the dynamics of the doctor–pregnant patient relationship.

Ultrasound was developed as an alternative means of viewing the fetus in the late 1950s, when the dangers of using X-rays to examine the fetus were finally recognised (Stewart et al. 1958; Oakley 1984). From improvised early applications, this low cost technology is now highly developed, in widespread use and provides a useful illustration of how scientific and technical uncertainties are managed in this medical speciality (Price 1990). A beam of very brief pulses of high-frequency sound, generated by a transducer placed in contact with a pregnant woman's skin, or more recently inserted vaginally, is directed through her body and partly reflected by soft tissues. The echoes from the reflections are visualised as an image on a screen for the operator to interpret.

Ultrasonography is presumed to be benign because no adverse effects have been demonstrated unambiguously in humans, yet many uncertainties surround the scanning procedure (Price 1990; Merritt 1989; Berkowitz 1993).[14] However, there has been relatively little discussion of the technology's potential for misuse, or how information about the ultrasound image is conveyed to the pregnant woman.

Ultrasound is used routinely in later pregnancy to allow clinicians to assess the growth and development of the fetus and to predict the delivery date. In fact, since the late 1970s, this has become the most routinised of the reproductive technologies in Britain, while commentators in North America have referred to the sonogram as 'a new pregnancy ritual'. Typically, women 'go for a scan' without signing consent forms and without any counselling. What they 'see', however, depends on the manner in which the image is presented to them.

Today, women undergoing a scan are allowed to participate in the scanning procedure in so far as they can view the image of their fetus. This was not always the case. Stuart Campbell's work at King's College Hospital in the early 1980s is often cited as the turning-point in this regard (Campbell et al. 1982). His research team demonstrated that pregnant women prefer to see, rather than not see, their fetus on the screen. And they prefer to be told something about what the scan reveals and to have various structures pointed out to them. Women who saw the scan of their fetus and who received information about it were deemed by the researchers to be more positive afterwards about both the scan and themselves.

On the basis of this work, the research team urged that women should be allowed to view scans. They surmised that fetuses would benefit from their mothers' more positive attitudes, and also reasoned that the women who were shown and told about their fetus on scan would be more likely to comply with health care recommendations, and, in particular, directives on smoking and drinking alcohol (Reading et al. 1982). It was at this point that the standard antenatal practice of limiting access to diagnostic images was abandoned in the interests of greater patient control and compliance.

But the scan remains, overwhelmingly, in the medical domain. The technique is performed in the interests of monitoring fetal growth and development to inform the clinical management of the pregnancy. A woman seeing her ultrasound scan is unlikely to be able unaided to interpret this way of rendering the world. In a study by Milne and Rich, most of the pregnant women undergoing ultrasonography reported considerable difficulty in identifying what they were seeing, although they recalled recognising movement and patterns. None challenged the authority of the scan operator in interpreting the image for them (Milne and Rich 1981).

There is evidence to suggest that most women who are offered an ultrasound scan regard it as more than a medical procedure, however. Hyde's study of pregnant women's attitudes to antenatal scanning indicated that women believed it would be interesting, reassuring or 'good to see the baby' even when there was no clinical indication to have a scan (Hyde 1986). When the scan is a routine antenatal procedure, women are encouraged to regard it as a positive process. They expect to be reassured. Indeed, in the clinic's appointment schedule, the brief time allowed for each scan assumes that 'all is well'. But these simple expectations are in practice easily disrupted. Quite apart from the uncertainties surrounding diagnosis by ultrasonography, the influence of the social organisation of the ultrasound technology has to be taken into account.

The division of labour in diagnostic ultrasonography is marked. The technology is operated by paramedical staff – radiographers, medical physics technicians or midwives – who are presumed to possess the skills needed to produce numerical and interpretational data. Usually, however, it is radiographers who report the findings to the referring clinicians (Witcombe and Radford 1986). The clinicians then use this data to make a diagnosis or prognosis concerning, for instance, the presence of abnormalities, multiple pregnancies, or projected date of delivery. If radiographers choose to communicate their findings to patients, to explain and to reassure, they are contravening the guidance given by the Disciplinary Committee of the Radiographer's Board in consultation with the Council for the Professions Supplementary to Medicine: 'No registered radiographer should knowingly disclose to any patient or to any unauthorised person the result of any investigation' (quoted in Witcombe and Radford 1986: 113).

This division of labour crucially interrupts the relationship between pregnant women and obstetricians. It is the radiographers who have to deal with the pregnant women together with the image on the screen. As data-gatherers, faced with the screen and the pregnant woman, they are not bound by the ethics of the doctor–patient relationship. Because of the restrictions on communication, the division of diagnostic responsibility between operator and consultant or GP may cause distress when an anomaly is found. Women watch the scanning procedure. They scan the faces of the operators intently. They observe their body language. They hear the exclamations, and emotion is a powerful communicative resource.

Women reported that they knew when 'something was not right', as did one woman who was later told that she was pregnant with triplets:

> They were scanning me up and down, you know. And I realised, you know, I said to myself 'How come they're taking such a long time? you know. Then the first nurse (scan operator) who was scanning me called another nurse (scan operator) over and you know *they were looking at the screen* and then *they were looking at each other* . . . they didn't say anything to me but they just kept scanning. And then, after a while, one of them turned

back, *looked at me and smiled.* So I said 'Is there anything wrong?' So she says 'No' and *carried on smiling* and screening me. *(Emphasis added)*

Describing her first ultrasound scan (which revealed to the operator that she might be pregnant with triplets) one woman explained that she had expected that going for a scan meant that she was 'going to come back with the results straight away'. But, as it turned out, she 'came back home with nothing', and spent the whole weekend worrying:

INTERVIEWER And did they detect three [fetuses] on the scan straight away?

MOTHER They did. But it was peculiar. They wouldn't tell me anything. Just wouldn't say anything. It was *dreadful.*

INTERVIEWER You saw the screen?

MOTHER Well I saw it. But it's not very easy to sort of pick out what's happening on the screen and she didn't volunteer any information. But I think it was awkward for her really. Because she couldn't tell me anything because it wasn't her place to. But she did, you know, to be fair, I think she *wanted* to tell me.

INTERVIEWER Can you describe what happened?

MOTHER Well I just said to her 'Can you tell me, you know, anything about the scan?' This was after she'd done the scan when I was dressed again and um she said 'No I'm sorry. I can't' she said. 'Your GP has to tell you that.' So she said 'But I will go and ring him for you if you want and see what he says.' Well, I mean, thinking back, I suppose common sense should have told me that it was nothing bad or she would have been – she didn't look upset or anything. Common sense would have told me that it couldn't have been all that bad or else – she didn't look upset or as if she'd found anything horrific there. But she also didn't give me any clues at all. And she went to ring him and he just said 'No! That she wasn't allowed to tell me. And so of course she came back and said 'You've got to make an appointment in a few days' time to go and see your doctor.' Because he thought he wanted the results of the scan and it would take a few days for them to reach, to go from the hospital to him. And so I had to wait a couple of days, which was *awful.*

INTERVIEWER Can you remember what went through your mind?

MOTHER Well I just thought that there was something wrong. I mean I just thought it was that there was something wrong with 'the baby'. I mean it just hadn't crossed my mind that it was twins. And I went back to work all upset about the whole thing. And my boss said 'Oh, I've got a friend and she's a midwife. I'll give her a ring and see what she's got to say.' So she gave her a ring and she just said 'Oh I shouldn't think there is anything to worry about' and she did say to my boss 'Oh I expect it is twins if they won't tell her' but she didn't tell me this because she didn't feel it was her place to you see. So I mean I just thought all the wrong things. I didn't think any of the right things. Just thought all of the wrong things I just thought that there was something wrong with 'the baby' you know.

Other women offered resistance: they reported that they became angry and refused to leave the clinic until their claim to be informed was recognised.

> The scan operator was very kind and calm at first and explained things to me as she went along until she had obviously found three babies. Then chaos! She would not tell me why at first. Four other scan operators came in and had a look. The radiographer at first told me a letter would be sent to my GP and for me to make an appointment to see the GP in 10–14 days time. This I could not accept and made a fuss until I was told there and then [that she was expecting triplets].

Their distress turned on the lack of acknowledgement of their need to know what had transpired. Another woman recounted:

> I was scanned by a radiologist at the first scan – told to come back the next day by a nursing officer. On asking why, was told couldn't tell me – come back tomorrow to be scanned by a doctor. I refused to move and was *then* told they suspected multiple pregnancy. Why I could not be told in the first place I do not know. It was, after all *my* pregnancy.

Lorraine Code has pointed to the ease with which self-doubt and dependence is created in such situations:

> The crucial difference, as I see it, turns upon acknowledgement. There is no more effective means of creating epistemic dependence than systematically withholding acknowledgement from a person's cognitive utterances; no more effective way of maintaining structures of epistemic privilege and vulnerability than evincing a persistent distrust in a person's efforts to claim cognitive authority; no surer way of demonstrating a refusal to know a person *as* a person than observing her 'objectively' without taking seriously what her experiences mean to her. (Code 1992).

The image on the screen is visibly separate from the pregnant woman and she is likely to become perturbed if her medical attendants respond entirely to the image and ignore her presence. One woman complained: 'the fact that I was even involved and upset seemed irrelevant'. Another woman reported that when her triplets were diagnosed, 'everyone rushed in to look at the monitor and ignored me'.

Pregnant with quadruplets, a woman described her long ordeal, as she was scanned and rescanned for two hours by one of the leading clinical specialists in obstetric ultrasonography:

> They had a better quality screen [than at her local hospital where she had had her first scans] so they pointed things out to me and explained certain things, but most of the time they talked to themselves with 'Ooohs!' and 'Aghs!' as if I was not even there! I was *occasionally* asked if I was O.K., but before I could answer properly they said 'Jolly good' and ignored me again.

In a survey conducted for the National Childbirth Trust (NCT) about its members' experience of ultrasound scanning a prominent theme in the letters received was of disquiet about the operator's lack of communicativeness or distant attitude. The report of the NCT survey included the quote: 'She ran the scanner over my stomach while discussing last Saturday's party with someone else' (Smith 1985).

Such is the belief in the power of the machine and its image as 'truth transporting', without any need for the woman as witness, that it can override a woman's own experience. One woman who was pregnant with triplets, for instance, became convinced that her babies were dead, even though she could feel kicks. She spoke about the experience of her recent scan:

> Unfortunately, on the last occasion I was scanned, I must have been down there almost three-quarters of an hour. And it is *very* distressing lying on the table. And she had to do several measurements twice because it was so difficult to get an accurate measurement, which again I understand because with three in there it is very, very hard for her [to measure].

She had felt very dizzy and queasy during the scan and had returned to her bed in a sideward of the maternity hospital, overwhelmed by the idea that, because she had not received reassurance from the scan operator, all was not well. Her conviction that this was the case was such that she discounted the fact that she could feel the babies kicking.

> Got back up onto the bed, [her husband] went, and I laid here for about three or four hours and I knew my notes had come up and the doctor walked in the room and flashed a tape measure round me and then shot off again. And I'm, of course, emotionally, emotionally you're at your lowest anyway and I'm convinced – you know: 'Oh my God, they're dead, they're dead.' I could feel them kicking, but you know, that was beside the point. And the nurses were avoiding all eye to eye contact, or so I thought. And by the end of the afternoon I was just in floods of tears because nobody had come and told me anything and I just thought: they can only be withholding information because it's bad news.

The deterministic status given to the imaging technique is a cause for concern (Daly 1989). Advances in ultrasound equipment yield not only more data, but also greater uncertainty about its interpretation: 'each machine update shows the observer features never recognised before' (Furness 1987). New features displayed on the scan have to be identified and this both extends the range of normality at each stage of gestation and adds to knowledge about human development. New observations, she cautions, contribute to an already sizeable list of visual anomalies – 'artifacts and red herrings' – which are sometimes distressingly misidentified as potential birth defects. Furthermore, some conditions previously identified only at birth can now be identified on scan, opening up clinical management decisions about the possibility of prenatal intervention. Nevertheless, a misdiagnosis or an intervention following diagnosis may lead to the termination of a wanted and normal pregnancy. Furness, an Australian radiographer, recounts how, 'an anxious nurse' cannot be dissuaded from the decision to terminate her pregnancy after she is told that she needs a repeat scan at 16 weeks, 'because her 12 and a half week fetus *may* have exomphalos' (Furness 1987). This is just one example of the cascade effect in clinical care where one medical intervention – the ultrasound scan – may lead to another with an unwanted outcome for the woman.

Conclusion

In reproductive medicine, images of fetuses and embryos are shown selectively to patients. Women were permitted to witness their ultrasound scans after research seemed to show that they 'liked' seeing the scan of their fetus and that the practice improved compliance in the double sense of greater social control of the patient, and greater legitimation of the technology. IVF clinics may also give their patients the opportunity to view their embryos as images projected from a microscope onto a screen. There the matter might well end as a topic of clinical concern. Yet dilemmas can arise when the patient also views the image, or can at least observe the observers, a witness to the image making.

When clinical practice involves collaborative tasks in the presence of women patients, they are well placed to 'see' (and conjecture about) what is 'going on' before they are told. More meets their eye than is on the agenda for discussion by their medical attendants. When all seems well, this may not appear a cause for concern: the mutual project on which both they and their doctors are engaged is reassurance. They are told what they have 'seen'. The control of information the women might wish to receive is, in medical practice, at the discretion of the clinician, as knowledgeable expert.

If patients perceive that something is amiss, this perception may contribute to a sense of anxiety that matters are out of personal control. Consequently, they may 'break the frame' and reinsert what has been relegated as 'beyond the frame'. This is difficult for the clinicians to manage, and their response may seem inappropriate. This perceived disjuncture in the management of uncertainty is highly significant in relation to discussions about the public understanding of science.

Similarly, measurements of the fetal image can confuse. Furness provides a striking example of a woman who was disturbed when information about her baby's development was conveyed, without explanation, in figures derived from measurements of femur and head circumference:

> A general practitioner rings to ask what he should tell Mrs Smith: she has just seen a report saying her fetus has a 32-week-size head and a 34-week femur, and she is convinced she has a long-legged mentally retarded child.

In such situations, the conditions of the transfer of knowledge become crucial. The responsibility for ensuring the pregnant woman sees and is told 'enough' to be reassured is in the hands of experts, who with benign intent may not only make her anxious but create what Garfinkel referred to as 'troubles': situations of disruption and confusion besetting unsuspecting people during the course of their participation in a routine procedure.[15]

The status structures and division of labour within medical practice are inimical in this situation. Communication and negotiation over knowledge needs is impeded. Women may be bewildered when what they witness is not explained, when the reasons for the refusal to provide information are not made plain and seem to deny accepted social norms about avoiding upsetting people unnecessarily. Their viewing is not contained by the frame of the imaging device. Trust may be undermined rather than enhanced. Women undergoing scans want to be reassured and readily sense when 'something is amiss'; as astute social observers, they pick up visual clues to uncertainty in scan operators and their colleagues. They may also glimpse the local uncertainties of the science which grounds the medical practice (Star 1985).

However, for both clinician and patient, the science-in-the-making underlying routine medical practice is normally obscured. In the clinic, the scanning device is 'black-boxed' and is in routine use. Moreover, the organisation of clinical practice does not allow for the social consequences of ambiguities or difficulties of interpretation. Similarly, in the IVF clinic, the uncertainties underlying the choosing of how many, and which, embryos to transfer is obscured by routinised practice.

In reproductive medicine, scientific models applied routinely in the clinic become locked up in medical practice, become established as expertise and become the 'fact' of the matter, fortified by the clinicians' professional status. Once established as medical expertise, clinical practice may be difficult to change. In this manner, medicine mediates, and obscures, the scientific uncertainties of ultrasonography and embryology and is sustained by professional structures characterised by closure and clinical autonomy and by the rhetoric of the doctor–patient relationship. These social processes produce premature social closure around still experimental and open-ended technical practices.

A diagnosis is only as good as the science underpinning the diagnostic test or device. By fostering belief in the efficacy of imaging devices and what images can signify, the medical profession risks both inducing unnecessary anxiety and creating unrealistic expectations about a standard of care that cannot be delivered.

The public image of science is also compromised. Science has enabled a privileged view, a 'window on the womb', and the development of promising diagnostic tests, but the contingencies of scientific evidence can get lost in the authoritative conventions of medical practice.

This is an important consideration in the public understanding of science as new reproductive and genetic technologies not only present new opportunities for intervention but influence the way people conceptualise the possibilities. The

geneticist Sydney Brenner has argued that the scientific discoveries of the Human
Genome Project can be conveyed unproblematically to the public through the
mediation of the medical profession.[16] He holds it best that the communication
of the 'new knowledge' about human genetic makeup be conveyed 'through the
conduit of current medical practice' (Brenner 1992). He argues that medicine
furnishes adequate protection; no novel 'ethical issues' arise. Yet clinicians are
responsible for conveying information in a way that manages uncertainty. What
they convey may not be 'good enough' to enable patients to appreciate the
uncertainties of the underlying science. Moreover, as long as medical practice
continues to be shielded from demands of democratic legitimation, biomedical
knowledge narrowly conceived is likely to be given primacy above any relational
knowledge (Beck 1992). This might not matter if it did not bear on certain
controversies and 'persuade' others into believing new 'facts' and behaving in
new ways (Miringoff 1991). Contemporary genetics has already profoundly altered
our notion of personhood. As Marilyn Strathern has reiterated, we need a
relational ground for our genetic knowledge (Strathern 1992).

It is important to focus on what is left out or obscured when medical practitioners
mediate reproductive and genetic knowledge through their practices, not least their
management of images. We need to question representational practices within the
clinic, to reveal how these practices are experienced, and what the consequences are,
particularly for women.

NOTES

1. The National Study of Triplets and Higher Order Births was the first population-based
   study of the problems faced by those responsible for the care of these children. I was
   responsible for the Parents' Study which was funded by the Department of Health (JS 240/
   85/13, see Price 1989).

2. Women and men attending designated infertility clinics were asked if they would co-operate
   in discussing what would be helpful written material to enable them to consider the prospect
   of a triplet pregnancy. These interviews were tape-recorded in the clinics. This project was
   also funded by the Department of Health and arose directly from the Parents' Study of the
   National Study of Triplets and Higher Order Births.

3. Lynch describes how these practices are accomplished by researchers working together in
   groups. He concludes 'The point is that the practices are necessary for constituting data,
   whether or not they are seen to be a source of error, and that it is only when they are taken
   for granted that the attribution of mathematical order to "nature" can succeed' (Lynch 1990:
   182).

4. As Pasveer illustrates, seeing by means of X-rays 'had to be learnt by doing'. She draws
   attention to the ways in which the specific content of X-ray images were *shaped* by the
   activities of early X-ray workers: 'there were no implicit meanings in the pictures'. And, as
   she remarks, 'The early X-ray images showed *too much* and therefore *too little*' (Pasveer 1990:
   365).

5. For example, artifact 'misrecognition' in early microscopy led to reports and illustrations of not only 'globules' in tissue structure, but also minute 'homunculi' in spermatozoa. Only later was the appearance of these phenomena attributed to aberrations in the lenses and to the imprecise optics of the microscopes in use at the time. See Bradbury, S. (1967a) *The Evolution of the Microscope* (Oxford: Pergamon Press); Bradbury, S. (1967b) *The Microscope, Past and Present* (Oxford: Pergamon Press); Lynch, M. (1985) *Art and Artifact in Laboratory Science* (London: Routledge and Kegan Paul).

6. Quantitative risk assessment is presumed to be free of institutional interests and constraints (Wynne 1987). Risks as probabilistic adverse outcomes (for which there is some measure of outcome magnitude) are distinguished from uncertainty. Probabilities about future 'states of the world' however, cannot be elicited where data are lacking, and risk problems are not well structured. This is the situation in relation to technologies in clinical practice and there is a paucity of research (O'Brien 1986: 41). It is more appropriate to refer to uncertainty rather than risk, given the level of contemporary disagreements about the definition, measurement, and comprehension of the probabilities to be attached to outcomes in medicine and also the complexities of institutional interests and constraints.

7. The 1990 issue of the IVF Congress Magazine published 'as a special information service to doctors' by one of the leading pharmaceutical companies in the field of IVF reported: 'Studies in Paris have shown that a combination of two non-invasive embryo tests do provide a possible prediction of pregnancy – first an embryo morphology assessment, and second, a chronological evaluation of embryonic growth in which the speed of cleavage is scored.' But then the article continued: 'While Dr Plachot [Dr Michelle Plachot of the Hopital Necker in Paris] agreed that the two tests in combination offered some predictive accuracy, she conceded that used alone each of the tests had produced conflicting and controversial results. Less controversial – but no less experimental – was a second non-invasive viability test in which embryonic health is measured by the embryo's ability to consume pyruvate. "The relationship between pyruvate uptake and embryo quality may provide the best assessment of viability" said Dr Plachot.' (Cited in item headed 'Embryo quality holds key to IVF's success' in *IVF Congress Magazine* 1990, p. 7).

8. Karen Dawson 'The First Fourteen Days' transcript of the BBC Horizon programme transmitted 26th February 1990, p. 16.

9. In the early 1980s it was authoritatively asserted that the IVF pregnancy rate would increase with the number of embryos transferred in each treatment cycle (Biggers 1981). By the mid-1980s pooled counts from IVF centres around the world served to confirm such predictions and encouraged the transfer of between three and six embryos in clinical practice (Seppala 1985). Concern about the marked increase in multiple pregnancies came later.

10. The annual statistics produced in the United Kingdom by the Interim Licensing Authority for Human In Vitro Fertilisation and Embryology (ILA, formerly the Voluntary Licensing Authority, VLA) show a clear association between the rise of multiple births from 1985 onward and the increased use of IVF, GIFT, and associated procedures (ILA 1991). Britain's Medical Research Council (MRC) Working Party on Children Conceived by In Vitro Fertilisation reported that 23 per cent of deliveries following assisted conception by IVF or GIFT occurring on or before 31st December 1987 resulted in a multiple birth of twins or more, compared with about 1 per cent for natural conceptions (MRC 1990). There is an additional risk factor: a higher-than-expected frequency of identical (monozygotic)

twins has also been observed, not only after the induction of ovulation with drugs, but also after IVF and GIFT (Edwards et al. 1986, Derom et al. 1987, Price 1989). Thus there are reports of three eggs or embryos being transferred in a GIFT or IVF procedure and the outcome being a quadruplet pregnancy.

11. The National Study showed that such children are, in all senses, high cost (Mugford 1990). Triplets and quadruplets are more likely than single babies to be of low birth weight and to be born prematurely, with all the associated neonatal difficulties and increased risk of disability and continuing developmental problems. Over half of the quadruplets and just over a quarter of the triplets weighed under 1500 grams at birth. Births occurred before 32 weeks' gestation in about half the quadruplet or higher order births and a quarter of the triplets. By contrast, information from the British Maternity Hospital In-Patient Enquiry indicates that only 1 per cent of singletons and fewer than 10 per cent of twins were born before 32 weeks. Furthermore, the National Study showed that 28 per cent of the triplets and 62 per cent of the quadruplets spent a month or more in neonatal intensive care. Mugford has calculated that the average National Health Service cost of hospital care is about £12 000 for triplets and over £25 000 for a set of quadruplets or more (Mugford 1990). Neonatal care accounts for more than 60 per cent of the total cost. And very few parents in the UK seek to pay for neonatal care in the private sector of medicine.

12. Hallam Medical Centre Patient Support Group Newsletter November 1990, p. 17.

13. Few people have known children from a triplet, quadruplet or higher order set. Even fewer have provided care or support for these children, for their parents, and for any siblings. For most people, the prospect of triplets or more is too remote to be imaginable. Ignorance seems understandable: the problems faced by those responsible for their delivery, care, and welfare are not widely known. When, however, clinicians advocate medical practices and procedures that increase the risk of such plural births, such ignorance becomes disconcerting.

14. The presumed 'safety' of ultrasound associated with diagnostic exposure is a sensitive and much discussed issue, particularly in relation to early pregnancy. However there is no clear-cut scientific advice because there is no scientifically adequate database for developing estimates of bioeffects and of risk. A number of reports have pointed to unresolved questions of risk associated with ultrasound exposure at levels generated by commercial diagnostic devices (Harris et al. 1989).

It is not the medical profession which has brought to the fore these issues about current knowledge and the different forms of practice, nor kept open the debate about the balance of possible benefits and hazards of routine ultrasound screening in obstetrics. In Britain it is the paramedicals, in the main radiologists, midwives, and consumer groups, who have taken the lead.

15. Garfinkel developed the use of 'troubles' as a discovery procedure by creating such situations (Garfinkel 1967: 35–75).

16. Royal Society meeting, London 1991.

REFERENCES

Anderson, D. C., 1987, 'Licensing work on IVF and related procedures', *Lancet* 1, 1373.
Arber, A., 1954, *The Mind and the Eye* (Cambridge University Press).

Armstrong, D., 1984, 'The patient's view', *Social Science in Medicine* 8, 737–44.

Beck, U. 1992, *Risk Society: towards a new modernity* (London: Sage).

Bennett, S. J., Parsons, J. H., Bolton, V. N., 1989, 'Two embryo transfer', *Lancet* 11, 215.

Berkowitz, R. L., 1993, 'Should every pregnant woman undergo ultrasonography?' *The New England Journal of Medicine* 329, 874–5.

Betteridge, K. J., 1981, 'An historical look at embryo transfer', *Journal of Reproduction and Fertility* 62, 1–13.

Biggers, J. D., 1981, 'In vitro fertilisation and embryo transfer in human beings', *New England Journal of Medicine* 34, 336–42.

Botting, B., Macfarlane, A., Price, F., 1990, *Three Four and More: a study of triplets and higher order births* (London: Her Majesty's Stationery Office).

Bradbury, S., 1967a, *The Evolution of the Microscope* (Oxford: Pergamon Press).

Brenner, S., 1992, 'That lonesome grail', *Nature* 358, 27–8.

Campbell, S., Reading, A. E., Cox, D. N., Sledmore, C. M., Mooney, R., Chudleigh, P., Beedle, J. and Ruddick, H., 1982, 'Ultrasound scanning in pregnancy: the short-term psychological effects of early real-time scans', *Journal of Psychosomatic Obstetric Gynaecology* 1, 57–61.

Code, L., 1992, 'The unicorn in the garden', *Women and Reason*, edited by E. D. Harvey and K. Okruhlik (Ann Arbor: University of Michigan Press) pp. 278–9.

Daly, J., 1989, 'Innocent murmurs: echocardiography and the diagnosis of cardiac normality', *Sociology of Health and Illness* 11, 99–116.

Daly, J. and Willis, E., 1989, 'Technological innovation and the labour process in health care'. *Social Science and Medicine* 11: 1149–1157.

Dawson, K. J., Rutherford, A. J., Margara, R. A., Winston, R. M. L., 1991, 'Reducing triplet pregnancies following in-vitro fertilisation', *Lancet* 337, 1543–4.

Derom, C., Derom, R., Vlietinck, R., Van Den Berge, H., Theiry, M., 1987, 'Increased monozygotic twinning after ovulation induction', *Lancet* 1, 1236–8.

Douglas, M., 1992, *Risk and Blame: essays in cultural theory* (London: Routledge).

Edwards, R. G., 1985, 'In-vitro fertilisation and embryo replacement: opening lecture', *Annuals of the New York Academy of Science* 442, 375–80.

Edwards, R. G., Mettler, L., Walters, D. E., 1986, 'Identical twins and in-vitro fertilisation', *Journal of In-Vitro Fertilisation and Embryo Transfer* 3, 114–17.

Edwards, R. G. and Steptoe, P., 1980, *A Matter of Life: the story of a medical breakthrough* (London: Hutchinson).

Foucault, M., 1979, *Discipline and Punish* (New York: Vintage Books).

Furness, M. E., 1987, 'Reporting obstetric ultrasound', *Lancet* 1, 675.

Garfinkel, H., 1967, *Studies in Ethnomethodology* (Prentice Hall).

Harris, G. R., Stewart, H. F., Leo, F. P., Sanders, R. C., 1989, 'Relationship between image quality and ultrasound exposure level in diagnostic US devices', *Radiology* 172, 313–7.

Hartshorne, G. M. and Edwards, R. G., 1991, 'Role of embryonic factors in implantation: recent developments', *Balliere's Clinical Obstetrics and Gynaecology: factors of importance for implantation'* edited by M. Seppala (London: Balliere Tindall).

Human Fertilisation and Embryology Authority (1991). *Code of Practice: explanation* (London: HFEA).

Hyde, B., 1986, 'An interview study of pregnant women's attitudes to ultrasound scanning', *Social Science and Medicine* 22, 586–92.

Interim Licensing Authority, 1991, *The Sixth Report of the Interim Licensing Authority for Human In Vitro Fertilisation and Embryology* (London: ILA).

Kassirer, J. P., 1992, 'Images in clinical medicine', *The New England Journal of Medicine* 326, 829–30.

Keller, E., Fox, and Grontkowski, C. R., 1983, 'The mind's eye', *Discovering Reality: feminist perspectives on epidemiology, metaphysics, methodology and the philosophy of science*, edited by S. Harding and M. Hintikka (Dordrecht: D. Reidel) pp.207–18.

Leese, H., 1981, 'An analysis of embryos by non-invasive methods', *Human Reproduction* 2, 37–40.

Leese, H. J., 1989, 'Energy metabolism of the preimplantation embryo' *Journal of Cell Biochemistry* 90, 188.

Leese, H. J., Humpherson, P. G., Hooper, M. A. K., Hardy, K., Handyside, A. H., 1990, 'Hypoxanthine guanine phosphoribosyl transferase (HGPRT) and adenine phosphoribosyl transferase (APRT) activities in single pre-implantation human embryos', *Biology of Reproduction* 42, supp. 1, 53.

Levene, M. I., 1986, 'Grand multiple pregnancies and demand for neonatal intensive care', *Lancet* 2, 347–8.

Levene, M. I., 1991, 'Assisted reproduction and its implications for paediatrics', *Archives of Diseases in Childhood* 66, 1–3.

Lynch, M., 1985, 'Discipline and the material form of image: an analysis of scientific visibility', *Social Studies of Science* 15, 37–66.

Lynch, M., 1990, 'The externalized retina: selection and mathematisation in the visual documentation of objects in the life sciences', edited by M. Lynch, and S. Woolgar *Representation in Scientific Practice*, (London: MIT Press) pp. 53–86.

Lynch, M. and Woolgar, S. (eds.), 1990, *Representation in Scientific Practice* (London: MIT Press).

Lyon, D., 1993, 'An electronic panopticon? A sociological critique of surveillance theory', *Sociological Review* 41, 653–78.

Martin, K. L., Hardy, K., Leese, H. J., 1990, 'Measurement of enzymes in single human pre-implantation embryos', *Journal of Reproductive Fertility*, abstract series 5, 58.

Merritt, C. R. B., 1989, 'Ultrasound safety: what are the issues?' *Radiology* 173, 304–6.

Milne, L. S. and Rich, O. J., 1981, 'Cognitive and affective aspects of the responses of pregnant women to ultrasonography,' *Maternal-Child Nursing Journal* 10, 15–39.

Miringoff, M. L. 1991, *The Social Costs of Genetic Welfare* (New Brunswick: Rutgers University Press).

MRC Working Party on Children Conceived by In-Vitro Fertilisation, 1990, 'Births in Great Britain resulting from assisted conceptions, 1978–87', *British Medical Journal* 300, 1229–33.

Mugford, M., 1990, 'The cost of a multiple birth', *Three, Four and More: a study of triplets and higher order births* edited by B. J. Botting, A. J. Macfarlane, and F. V. Price (London: HMSO).

Oakley, A., 1984, *The Captured Womb: a history of the medical care of pregnant women* (Oxford: Basil Blackwell).

O'Brien, B., 1986, *What Are My Chances Doctor? a review of clinical risks* (London: Office of Health Economics).

Pasveer, B., 1990, 'Knowledge of shadows: the introduction of X-ray images in medicine', *Sociology of Health and Illness* 11.

Pateman, C., 1992, 'Equality, difference, subordination: the politics of motherhood and women's citizenship', *Beyond Equality and Difference: citizenship, feminist politics and female subjectivity*, edited by G. Bock and S. James (London: Routledge).

Petchesky, R. P., 1987, 'Foetal images: the power of visual culture in the politics of reproduction' *Reproductive Technologies: gender, motherhood and medicine*, edited by M. Stanworth (Cambridge: Polity Press).

Peters, H., Nervell, S. J., Obhrai, M., 1991, 'Impact of assisted reproduction on neonatal care', *Lancet* 337, 797.

Pfeffer, N., 1987, 'Artificial insemination, in-vitro fertilisation and the stigma of infertility' *Reproductive Technologies: gender, motherhood and medicine*, edited by M. Stanworth (Cambridge: Polity Press).

Pfeffer, N. and Quick, A., 1988, *Infertility Services – a Desperate Case* (London: Greater London Association of Community Health Councils).

Price, F., 1989, *The Parents' Study: Final Report to the Department of Health* (London, Department of Health Library).

Price, F., 1990, 'The management of uncertainty in obstetric practice: ultrasonography, in-vitro fertilisation and embryo transfer', *The New Reproductive Technologies* edited by M. McNeil, I., Varcoe, and S. Yearley (London: Macmillan).

Price, F., 1991, *The Prospect of Triplets or Quads: a project to pilot an information booklet for women and men attending infertility clinics.* Final Report to Department of Health, London, Department of Health Library.

Price, F., 1992, 'Isn't she coping well?' Providing for mothers of triplets, quadruplets and quintuplets', *Women's Health Matters*, edited by H. Roberts. (London: Routledge).

Reading, A. E., Campbell, S., Cox, D. N., Sledmore, C. M., 1982, 'Health beliefs and health care behaviour in pregnancy', *Psychological Medicine* 12, 379–83.

Reiser, S. J., 1978, *Medicine and the Reign of Technology.* (Cambridge University Press).

Rorty, R. (1979) *Philosophy and the Mirror of Nature* (Princeton University Press).

Scott-Jupp, R., Field, D. M., Macfadyen, U., 1991, 'Multiple pregnancies resulting from assisted conception: burden on neonatal units', *British Medical Journal* 302, 1079.

Seppala, M., 1985, 'The world collaborative report of in-vitro fertilisation and embryo replacement; current state of the art in 1984', in Seppala, M., Edwards, R. G., (eds.). 'In-Vitro Fertilisation and Embryo Transfer', *Annuals of the New York Academy of Science* 442, 558–563.

Seppala, M. (ed.) 1991, *Balliere's Clinical Obstetrics and Gynaecology. Factors of importance for implantation* (London: Balliere Tindall).

Sidaway v Gov of Bethlem Royal Hospital (HL(E)) *The Weekly Law Reports* (8 March 1986) 480–502.

Smith, B., 1985, 'NCT ultrasound survey results', *New Generation* 4, 5–6.

Staessen, C., Janssenswillen C., Devroey P., and Van Steirteghem, A. C., 1993, 'The replacement of two "good-quality" embryos does not reduce the conception rate", *Human Reproduction* 8 (supplement 1), abstract 138.

Star, S. L., 1985, 'Scientific work and uncertainty', *Social Studies of Science* 15, 391–427.

Steirteghem, A. C., 1994, 'One year's experience with elective transfer of two good-quality embryos in the human IVF and ICSI programmes', Abstracts of the 10th Annual Meeting of the ESHRE, Brussels, 29 June 1994.

Stewart, A., Webb, J., and Hewitt, D., 1958, 'A survey of childhood malignancies', *British Medical Journal* 1495–508.

Stewart, N., 1986, 'Women's views of ultrasonography in obstetrics', *Birth* 13 (special supplement) 34–7.

Strathern, M., 1992, *Reproducing the Future: anthropology, kinship and the new reproductive technologies* (Manchester University Press).

Strong, C., 1987, 'Ethical conflicts between mother and fetus in obstetrics', *Clinics in Perinatology* 14, 313–21.

Turner, K., Martin, K. L., Woodward, B. J., Lenton, E. A., and Leese, H. J., 1994, 'Comparison of pyruvate uptake by embryos derived from conception and non-conception natural cycles', *Human Reproduction* 9, 2362–6.

Waterstone, J., Parsons, J., Bolton, V., 1991, 'Elective transfer of two embryos', *Lancet* 337, 975–6.

Westmore, A., 1984, 'History', *Clinical In Vitro Fertilisation*, edited by A. Trounson and C. Wood, (Berlin: Springer-Verlag).

Willis, E., 1983, *Medical Dominance* (Sydney: George Allen and Unwin).

Witcombe, J. B. and Radford, A., 1986, 'Obstetric ultrasonography: wider role for radiographers?' *British Medical Journal* 292, 113–55.

World Collaborative Report, 1993, 8th World Congress on IVF and AAR, Kyoto, Japan.

Wynne, B., 1987, *Risk Management and Hazardous Waste: implementation and the dialectics of credibility* (Berlin: Springer-Verlag).

Wynne, B., 1991, 'Public perception and communication of risks: what do we know?' *Journal of National Institutes of Health Research* 3, 65–70.

Yoxen, E., 1988, 'Public concern and the steering of science', Report for the Science Policy Support Group. April.

5 Ignoring science: discourses of ignorance in the public understanding of science

MIKE MICHAEL

Introduction

The starting-point of this chapter is the observation that people do not simply possess knowledge about scientific 'facts' and scientific procedures and processes, they can also reflect upon the epistemological status of that knowledge. In addition, I argue that this active reflection can directly affect their responses to science and scientific experts. In feeling uncertainty about their understanding of science, or in identifying a 'lack' in their knowledge, people are making tacit judgements in relation to the authoritative source or sources of that knowledge. Thus people can review the standing of their scientific and technological knowledge in relation to some more or less expert source such as scientists, the media, friends and relatives, and so on. As such, identity cannot but be implicated. Conversely, the ways in which people regard themselves and the value they place upon their scientific knowledge, affects the ways in which they understand science. We have, therefore, a sort of discursive jigsaw in which identity, the status of lay scientific knowledge, and scientific expertise are delimited.

This chapter will, first of all, briefly examine three approaches to the public understanding of science which are essentially interested in describing the scientific knowledge that people 'possess'. The fact that these approaches ignore issues concerning the reflexivity and identity of lay people, suggests their underpinning model of the individual is fundamentally mechanistic. The implication is that a particular representation or narrative of the lay person is promoted; in some cases, this is further disseminated through the media. These 'knowledge description' approaches will then be contrasted with a social constructionist/discourse analysis of the public understanding of science. Further, I will focus on one particular facet of the public understanding of science. When talking to lay people about their scientific knowledge, in many cases we found that people simply do not possess any of the 'relevant' (at least for the investigator) scientific knowledge; they do not simply have a 'defective' body of quasi-scientific knowledge, they have none at all. However, people are very adept at reflecting upon this manifest absence of knowledge, that is, upon their 'ignorance'.[1] The aim of this paper is to explore the discourses of ignorance that people mobilise when reflexively commenting upon their lack of scientific knowledge. We will take it that people are not merely 'rationalising' to save face; their reflections also represent other social reasons for defending their 'ignorance', reasons that overspill the confines of the immediate situation. Thus, our task is to trace some of the functions that these discourses might fulfil in formulating a social relationship between themselves and science. We take it as given that the discursive route that people take in locating themselves and their 'ignorance' in relation to science reflects and mediates the sorts of social identities available to them. Such identities will range from the general (what it is to be a member of the lay public) to the local (what it is to be a member of a particular community, profession, or social group). However, while I will make suggestions as to what sort of social identities are being mobilised and realised through discourses of ignorance, my present purpose is more modest: namely to map out these discourses and the ways in which these contrive a relation between self and science.

Knowledge description approaches

The three approaches I will consider here are: survey analyses of the contents of the public understanding of science and of attitudes towards science; the theory of social representations; and the 'mental models' approach.

The first approach to public understanding of science to be considered here is that entailing survey methodology. As the Royal Society report[2] notes:

> There are many surveys of attitudes towards science and technology both in the UK and overseas, especially in the USA. But there has been much less effort outside the formal education system devoted to assessing the understanding of science and technology.[3]

Where comparisons are drawn the links are not at all clear. Beveridge and Rudell[4] in reviewing the 1985 Science Indicators Report note that, while expressions of interest in science are great, science informedness is considerably lower (irrespective of whether such informedness was self-reported or externally measured). These observations have received further support from survey work carried out under the Public Understanding of Science Programme by John Durant's team.[5] The Royal Society's lament that the public lacks scientific knowledge appears, in light of such evidence, increasingly worrying.

What assumptions are operating here? Clearly, we see a neglect of the reflexive and social character of understanding. To a somewhat greater extent than the 'mental models' and social representation (see below) approaches, people are treated as essentially repositories of information. In all three cases, the analyst dips into them, and, with the relevant methodology, resurfaces with a description of their contents, that is, the 'understandings' of science. This reductionist apprehension of the layperson and her knowledge is, as will be elaborated below, highly problematic and misleading. The 'mental models' and social representation approaches tend to define (and even celebrate) the contents of these repositories as eccentric: the extent to which these understandings deviate from the norms of science are explained in terms of cognitive or practical functions. In contrast, the survey research tends to focus upon the intellectual deficiencies in people's understandings as measured against some objective or authoritative body of scientific knowledge. Indeed, it adheres to what Brian Wynne has dubbed the *Deficit model*.[6]

This view of the public as mechanistic and deficient with regard to the understanding of science is not confined to the academic domain.[7] The media publicity given to the findings of such survey work has been considerable. With sub-headlines such as 'With more than a third of the population not knowing that the earth goes round the sun, Britain could be in serious trouble' (*Sunday Times*, 19 November 1989), the narrative of public deficit is conveyed to a wider audience, and the contrast between a knowledgeable science and an ignorant public is reiterated.

However, the survey research is informed by a specific policy purpose: to encourage and nurture democratic participation. Only by identifying and measuring the gaps in people's understanding can the level of scientific literacy be raised to that required to make informed judgements in contemporary democracies. Yet the remedies for this 'deficit' merely reinvoke the power-relation of a dominant science and a subservient public. Thus the Royal Society recommends that there should be an increase in the amount and quality of science education, in media coverage of science, and in scientists' popularising input into the public sphere: science is the active disseminator and the fountain of meaning and agency, the public are merely the passive receivers and repositories.

This contrasts to the second 'knowledge description' approach that I will consider – the social psychological perspective known as social representation theory. Moscovici[8] has characterised social representations as concepts, statements, images, and explanations that originate in the course of inter-individual communication and whose prime function is to make the unfamiliar familiar. The particular relevance of this perspective on research into the public understanding of science is that social representations are considered to render the unfamiliar productions of science familiar. Surprisingly, the range of social representations that have been studied includes relatively few drawn from the sciences. For example, from the biological sciences social representations of Aids[9] and health and illness in general[10] have been studied; from the social sciences public opinion polls[11] have been considered.

One study that has engaged with physical processes is that conducted, two weeks after Chernobyl, by Galli and Nigro,[12] into social representations of radioactivity among Italian children aged between nine and twelve. These workers found that children's drawings of radiation often took the form of a dark pink cloud. This image could be traced back to representations of the Chernobyl cloud that had appeared on television. Despite the source being the media, Galli and Nigro concluded that for the majority of children who had (re)produced the 'pink cloud' representation, the radioactivity *was* the pink cloud. Chernobyl radioactivity had, in Moscovici's terms, become 'objectified' as a pink cloud.

It would seem unwise however to generalise this link to other situations and groups. For example, adults might well treat such representations as just that – concocted media representations of a complex physical phenomenon. They would recognise that, while they themselves might not be able to reformulate the radiation in terms other than a pink cloud, radiation would nevertheless still be distinct from its media representation. This raises the vital issue of the reflexive assessment of the wider credibility of particular social representations. What is missing in Galli and Nigro's work is an analysis of the ways that people might construe the epistemological status of their knowledge. This would include, crucially, a broader consideration of how, in those cases where knowledge is 'lacking' or where there is an unwillingness to reveal what is known, laypeople construct this absence. This would tap into the sorts of relationship between the layperson and science that are implied in reflection upon 'ignorance'.

Finally, the third knowledge description approach is that developed by cognitive psychologists who have examined the understanding of science and technology in terms of the 'mental models' that people have of numerous scientific or technological phenomena. Thus we have studies of: motion,[13] electricity,[14] home heat control,[15] and evaporation.[16] Quinn and Holland[17] have suggested that 'mental models' of physical processes can be picked up and put down at will: they serve as tools. The emphasis is thus on the instrumental use of such models and as such there is little examination of the social and cultural contexts of these models. This stands in contrast to those studies of cultural models of social phenomena such as marriage,[18] gender types,[19] and mind[20] which do engage with the social embeddedness of such cognitions.

The main point I want to raise in relation to the appropriation of public understanding through the academic construct of 'mental models' concerns 'familiarity'. If we are to conceptualise 'mental models' as 'tools', we would also expect that expertise in their use would be dependent on some sort of practice, and incremental familiarity, with the relevant knowledge domain. The studies mentioned above deal with respondents who have a familiarity with, and sometimes a profound practical understanding of, the knowledge domains being studied by the investigator. In those more typical cases in which people have only sporadic or fragmented contact with the relevant knowledge domain, it is difficult to see how 'mental models' could have been derived. Thus, in our research, when we attempted to get at the 'mental models' of ionising radiation by using the standard technique of posing various puzzles (for example, what is more dangerous, low prolonged doses of radiation or short high doses?), what we found again and again was a peremptory attempt to answer appropriately followed by a comment on the status of that answer. In other words, despite people's initial willingness to play the game, they found they could not do so convincingly and resorted to glossing that 'inability'. As such they were shifting from 'helpful respondents' to 'reflexive critical commentators' and in the process drawing upon a much broader context and array of social identities.

This brings us to another dimension which the mental-models approach neglects: the morality of knowledge. The knowledge domains themselves are not socially neutral: they signify a concomitant institutional and moral frame. This concerns the perceived 'right' that one has to talk about ionising radiation. The instances of 'mental models' outlined above – motion, electricity, home heat control, and evaporation – not only have greater familiarity (in general and to the specific respondents in these studies), they also have a different institutional and political resonance. It is reasonable to expect, given (some) people's social positioning, that they would be more comfortable musing on motion or home heating systems than on ionising radiation. In other words, given participants' social uncertainty a propos ionising radiation, it is unsurprising that there is easy recourse to discursive forms which limit the epistemological status of their knowledge. But, simultaneously, this is not simply epistemological circumspection, as we shall see, it is a political process whereby our participants construct a more or less critical relation with science and the institutions of science.

In sum, there are three problems with the knowledge description approaches outlined above. Firstly, ignorance cannot be treated as simple deficit: it entails active construction. Secondly, in the process of that construction, people reflect upon the epistemological status of their knowledge. Thirdly, in the act of such reflection, social and political contexts are drawn upon in order to resource a relation to science. So, the central thread of this chapter is that people can reflect upon their relations to scientific knowledge and/or its institutional embodiments. As such they draw upon discourses that address the differences in knowledgeability. The general point is that lay understandings of science and scientists are supplemented by people's recognition of their 'ignorance' of the actual workings of science. Therefore, a necessary, but so far neglected, component in the public understanding of science concerns the ways that people formulate this relation of difference from science – a difference that focuses upon one's own 'ignorance' of its workings. On the level of methodology, such a focus on 'ignorance' offers an important means of probing the public understanding of science. On the substantive, theoretical level, we can examine how lay people's self-ascriptions (of ignorance) set up a relationship between themselves and science – one in which the causes or reasons behind their 'ignorance' reflect their social identity and their relations of dependence, co-operation, or challenge – that is relations of power *vis-à-vis* science.

This approach therefore explores the ways that people actively construct, and defend, their absence of knowledge through what I will call *discourses of ignorance*. This is not only an interesting research question, it has important educational and policy implications. Facing any programme of science knowledge dissemination is the underlying barrier represented by these discourses. Discourses of ignorance not only constitute a means of understanding and explaining one's lack of knowledge, they also signify and reflect the perceived social relations between science and lay person, between self and expert. As such, they are connected with questions of social or cultural identity, and any educational programme must address what the knowledge it disseminates implies for these socially embedded conceptions of self and other. As mentioned above, I will make some suggestions as to the types of identities that feature in the uses of discourses of ignorance, however, my immediate aim is to document these. A more rounded analysis of these social identities would need to study the social and cultural backdrop in which the speakers operate, either through ethnographic study (of the kind found, for example, in Chapter Six) or by imputing a certain cultural commonality (as suggested in Chapters One and Two).

Ignorance

In considering the way that public 'ignorance' is constructed, I will depart from Ravetz's treatment of 'ignorance' in the science policy field.[21] Ravetz's concern is with the effects of 'ignorance' in the process of decision-making – 'A decision problem involves "ignorance" when some components which are real and significant are unknown to the decider at the crucial moment.'[22] Ravetz goes on to suggest that such 'ignorance' is only discovered retrospectively by decision-makers, and that it is inherent in the application of technological systems. In contrast to this conception of 'ignorance', Smithson defines 'ignorance' in the following way: 'A is "ignorant" from B's viewpoint if A fails to agree with or show awareness of ideas which B defines as either actually or potentially factually valid.'[23] This subjectivist formulation avoids 'confounding judgements by the social scientist about the validity of the cognition being studied'.[24] Smithson duly notes that A and B can be one and the same person in so far as an individual can ascribe 'ignorance' to self. Smithson's concern is to treat 'ignorance' as a serious topic for social theory and, to this end, he presents an impressive array of instances of social life in which 'ignorance' plays a pivotal role. For example, he considers: norms against knowing (for example, as evidenced in politeness phenomena); 'ignorance' strategies and games (for example, in the courtroom questioning of expert knowledge); the particular settings and occasions of 'ignorance' (for example, confessions); the role, scripts, and identities associated with 'ignorance' (for example, the requirement of selective inattention or 'ignorance' for the successful performance of specific duties). Smithson is primarily interested in surveying and demarcating the multiplicity of social contexts in which 'ignorance' is constructed, deployed, and ascribed. In contrast, the present chapter's more modest ambition is to take up just one strand from Smithson's subtle and wide-ranging survey. That is, I will explore the way that people use particular discourses to reflexively comment upon manifest 'ignorance' (in relation to science and ionising radiation), and thereby tentatively to position themselves in relation to the relevant institutions and groups, both expert and lay.

Notes on method and analysis

Adopting the methodological approach of discourse analysis,[25] I examine the ways that respondents construct, through their talk, a particular vision of their apparent lack of scientific knowledge and thereby a view of themselves-in-relation-to science. In addition, the present work parallels the analysis of Wetherell and Potter.[26] These authors considered the use of narrative characters – discursive constructions of self – in the mitigation of actions that were possibly blameworthy. They analysed the talk of people who mainly excused or justified violent police behaviour towards anti-apartheid demonstrators. While there are similarities in the responses given by Wetherell and Potter's interviewees and ours – actions were characterised in terms of rational motivation, role requirement, and natural disposition – I am not centrally interested in examining these discursive forms in terms of their mitigating function. Rather, the present emphasis will be on the way that people use these discursive devices to construct a relation to science and scientific knowledge – a relation which, I suggest, entails a vision of dependency and autonomy.

While there is inevitably an element of social accountability present in the use of discourses of ignorance, I will concentrate on the projected relation with science that this talk constructs. In other words, I do not treat the talk of the respondents as directed to the interviewer alone; rather, the talk also, quasi-conversationally and over the interviewer's head so to speak, addresses science or scientists. That is, we can read the talk of respondents as containing elements of a negotiation with science in which the relative knowledge levels, the right to speak about science and the relevance of particular knowledge domains is, in a highly compressed way, articulated. To put it another way, people are defining domains of authority for different kinds of knowledge in relation to themselves, hence articulating (at least in part) social identity. More generally, we are here treating the respondents' talk as involved in a range of conversations and whose 'interlocutors' include the interviewer, science and scientists of various sorts, 'in-groups', and so on.[27] So, the present analysis de-emphasizes the immediate face-to-face situation, and investigates the meanings that such talk has with respect to the relation between the speaker and wider, physically absent audiences (for example, scientists).

To illustrate and explore some of these social meanings of discourses of ignorance, I draw upon three pieces of fieldwork. Semi-structured interviews about ionising radiation were conducted in the following contexts:

A   Volunteers in a Radon survey carried out by a Local Council Environmental Health Office. Volunteers kept a small plastic Radon detector in their homes for six months. They were also provided with a sheet of details on Radon and the rationale behind the survey. Twenty-four interviews were conducted with self-selected volunteers several months after the detectors had been removed for analysis.

B   Interview Panel: Initially members of the public were chosen at random from the electoral register. This led to some snowballing. Several panel members were recruited from other groups in our research programme such as the Radon Survey volunteers and patients attending a local hospital for X-rays. The aim of the panel was to conduct repeat interviews and, depending upon time of original contact and availability, panel members participated in from 4 to 1 interviews. The interviews covered not only the understanding of ionising radiation, but also of information technology and ultrasound, and the perception of science and scientists in general.

C   About 20 time-served electricians working at the Sellafield Reprocessing Plant interviewed *en masse* while attending evening classes at a local college of further education.

I make no attempt to quantify the rates or proportions of these various types of response. The present chapter simply aims to sketch out an array of discourses in order better to apprehend how it is people address, bracket, and articulate issues of science and technology. It is, then, perhaps prudent to regard the discourses that will be outlined below as ideal types.[28]

Before describing a number of discourses of ignorance I want first to briefly recap on an important proviso. Although my analysis aims to unpick some of the discourses that people can draw upon in order to construct a relation to science, these discourses are manifested in very specific circumstances, namely the interview situation. Here, people engage in what can be called *focused articulation*. Obviously, the interview is about science, ionising radiation, and expert knowledge; people are asked to formulate understandings of these things. It is thus important to place the centrality of these issues in the appropriate wider context. I have already mentioned that in formulating their relationship to science, lay people will be drawing on an array of identities. Ethnographic work is necessary (see various chapters in this volume) to locate such situated accounts, articulations, and formulations in the broader cultural landscape and to detail how these interact with the multiplicity of other social relations in which people are involved. Nevertheless, at the very least, the present analysis gives a picture of the range of discursive resources that people can draw upon in this context, and foregrounds the reflexive dimension that can inform people's overt responses.

Discourses of ignorance

A Unconstructed absence

First, we can note that often the fact that people were manifestly lacking in specific types of knowledge did not occasion further commentary from them. That is, no attempt is made to discursively construct absence. Thus, in many cases, even where they would appear to have had some direct interest, in response to questions about ionising radiation, people would simply answer 'don't know' or 'never thought about it'. Here, it seems that absence was only brought to people's attention by virtue of the question, and that the absence of an answer was simply a fact of life, unworthy of elaboration or explanation. In a way people treated absence as an *irrelevance* or peripheral to their primary concerns. However, in many cases, a 'don't know' would be followed by a request for the correct answer: 'Well what does Radon do?' or 'What are alpha particles?' This suggests that the deficit or absence is recognised and the interviewer used as a potential source of information. Absence here is linked to the opportunity to rectify; indeed, the subsequent question is used to preempt an impression of uninterest, lack of curiosity, indolence, or whatever. The implication is that for some unspecified reason or circumstance the interviewee has been unable to seek out this information. Here, then, the exposed ignorance does have some significance for the respondent. Absence is implicitly demarcated in this instance by use of requests for information; these in turn signify that the self is, at the very least, an interested and responsible 'member of the public'. In the following examples, I will look at the way that absence is explicitly constructed.

B 'Ignorance' and mental constitution

In some cases, absence is recognised as an entrenched, global scientific 'ignorance', as a 'not-knowing', a lack of education. It might be the case that this is used as a defensive strategy in the course of the interview to avoid responsibility for specific perceived mistakes. For example:

(A)

I: Any idea why it [Radon] should bubble up, rather than things like carbon dioxide . . .

AW: Probably a light gas, just a gas . . .

I: I'm not sure it does actually . . . [Laughter].

AW: Oh yeah . . . My knowledge of science is zero . . .

(Radon: AW. 31)

Later on this participant, a museum director, goes on to describe how he thinks Radon is detected through the magnification of fission tracks in perspex contained in the monitoring devices. He mentions that he knows about fission track data analysis through archaeology – yet his 'knowledge of science is zero'. It seems to me that here there is a bracketing of scientific knowledge as 'other'. The knowledge that this speaker possesses cannot conceivably (in his own estimation) be science. It might be the case that he considers his knowledge insufficiently detailed or arcane to count as science (and perhaps here he is drawing parallels with the criteria of archaeological expertise which reflect similar rigour, specialism, and so on). Thus, the criteria of what counts as scientific knowledge are so strict that he is not able to aspire to them. However, in this instance there is only an indirect ascription to self. In contrast, other speakers' meta-commentaries upon their knowledge entail a characterisation of self, namely the lack of scientific mind. Thus:

(B)

I: Do you think of it [Radon] as er . : . as er rays or gas?

LG: Erm . . . gas . . . probably . . .

I: Any idea how it gets produced in the ground?

LG: No idea . . . [laughs]

I: Have a guess.

LG: Erm . . . I'm not very scientifically minded . . . I don't know . . .

(Radon: LG. 25)

(C)

I: Any idea what the processes are by which this gas is created?

PC: No . . . I would imagine it's just a gas created by the natural breakdown of the various components of the soil. I mean that's just an assumption really not . . . I'm not very scientifically minded . . . I know there are these things happening in the earth . . .

(Radon: PC. 28)

(D)

    I: Do you know what they were . . . measuring?

MC: Radioactivity weren't they?

    I: Do you know what sort?

MC: No . . .

    I: You don't and they didn't say . . .

MC: I didn't know whether they . . . as I say I'm not science minded . . . technically minded, anything like that . . . but my husband works at Great Lakes Chemical Plant . . .

(Radon: MC. 49)

These ontological statements about the self are similar to what Halliday[29] has called in his functional grammar, relational processes that describe processes of being. These are either intensive and take the form of such statements as 'I am not scientifically minded', or possessive such as the sentence: 'I don't have a scientific mind.' In the present context, such responses constitute meta-commentaries or reflections that cast doubt upon the substantive statements made about Radon and ionising radiation: as such, they problematise the speaker's right to address these scientific topics.

Here, the speakers are not merely differentiating themselves from science *per se*. In addition, they are saying that they are constitutionally not mentally equipped to fathom the mysteries of science – whether those be the domain of professional scientists or the domain of members of a public 'scientific culture'. There is, then, a genetic differentiation between self and science, and, further, in the context of the interview, speakers using this discourse situate themselves in a quasi-dependent position relative to science. Nevertheless, in stressing one's lack of scientific mind, one can also point to one's own positive attributes or functions. In the present context, this contrast is neatly encapsulated by a Lancaster Town Hall receptionist (Radon Survey volunteer), who noted, when talking about scientists in the guise of Lancaster City Council Environmental Health Officers: 'They couldn't do my job and I couldn't do theirs.' This discourse will be elaborated below.

The relation of dependence to science is partly born of exclusion from it. While the constitutional ignorance discourse formulates this in terms of the speaker's own (in)capacities, in other cases exclusion can be ascribed to the nature of science. Thus, L, a member of our interview panel in her early forties, when discussing a questionnaire question on the science-ness of various disciplines remarks: 'it's more of a science because I know nothing about it. Physics and Chemistry, yes, it's a science because I know nothing about it'. Here, the less insight she has into a subject, the more science-like it is. At the very least this statement reflects her perceived status as an outsider. This status reappears in the reflective point that she makes: 'you see you can also look at it [science] as to whether it's relevant and important and how important it is depending on how scientific it is . . . [laughs]'. In this instance, L sets up a circularity: science is important, it is science because it is important and so on; science is being represented as a self-defining enterprise. Once again she is situated on the outside of this hermetic world. Here, rather than simply noting one's lack of scientific knowledge or attributing it to one's lack of scientific mind, the cause for one's distance from science is partially to be found in the nature of science itself: it is the hermetic, self-defining form of science that serves to exclude L.

c *'Ignorance' and the division of labour*

In addition to the constitutional basis for absence, there is also a functional view of the difference between science and other spheres of action or expertise. Absence can here be put down to the *division of labour*. I will present two examples of this.

(E)

> **I:** What's strange was when I started reading all this stuff I was dumbfounded about how much I didn't know . . . really . . . and a lot of it is still sort of dodgy especially the biological stuff. . . .
>
> **DM:** Well, I read quite a bit of it but . . . but it's not my job.
>
> *(Panel: DM. 1.72)*

The speaker in extract (E) is a statistician. He is denying the necessity to know these things because it is not his job. This account straddles the duality of structure and agency – of the requirements imposed by the social order on the one hand, and personal volition on the other. There is here a dual rhetorical strategy to explain absence: firstly, in terms of external conditions – I am *not required* to know this stuff; secondly, in terms of internal, personal disposition – *I don't need* to know this stuff. Thus the ambiguity of the 'role' construct, as simultaneously implying agency (or volition) and social structure (or necessity), can be discursively exploited.

A similar strategy is instanced in the following exchanges between ourselves and some twenty Sellafield time-served electricians. Unfortunately, conditions meant that we could not record so the following is not guaranteed verbatim. What follows is derived from field notes written immediately afterwards. The following talk occurred in the context of describing what they, as electricians in the course of the daily work procedures, needed to know about radiation. One electrician noted that: 'we're not employed to do what the monitors do . . .'. The sole woman electrician followed this with: 'People [i.e. the electricians and possibly other blue-collar workers] don't have to know too much, you've got to trust someone somewhere, and they're [the health physicists] trained for it.' Later, she added: 'If people knew too much, they would panic in an emergency because they know just how dangerous it really was.'

In the former quotes, absence is formulated negatively: the absence of knowledge is compensated for by other functionaries – specialists who possess the requisite knowledge and use it for the benefit of the electricians. But the absence is also formulated positively in that it serves to preclude panic under certain circumstances. It is a good, functional thing to remain ignorant for the sake of being clear about proper procedures.

Now this contrast and complementarity is evoked in the very particular circumstances of the nuclear industry. However, in a much less articulated way, people seem to hold to the same division of labour view of their relation to science. (That it is less articulated is hardly surprising given people's relatively impoverished contact with scientists.) For example, in our Radon Survey sample, the complementarity between scientists (environmental health officers) and volunteers was framed around the common goal of 'doing good'. That is to say, the scientists (with whom the volunteers had little contact) and the interviewees-cum-volunteers were functioning together (I hesitate to say working together) to conduct a survey of Radon levels in Lancaster and its environs for the general good. Thus the common goal bound scientist and layperson together as a collaborative unit; the function of the volunteer was as a sort of guinea pig or perhaps as concerned citizen – there was no reason to go beyond this in the sense of finding out about Radon and its properties even if there was a Radon measuring device sitting in one's living-room for six months. Superficially, the measuring device might appear to be an irresistible stimulus to go and find out more about Radon, even if the knowledge gained was at best elementary. However, in the context of a division of labour, the device came to signify or symbolize one's passive role as volunteer – it spurred the volunteer actually not to find out. Indeed, it was remarkable how many of the volunteers, though they expressed to us an interest in finding out more, in practical terms made little if any effort to seek out information on Radon and the issues surrounding it.[30]

The positive formulation of absence is also present in the final broad discourse of ignorance: here, rather than attribute absence to constitutional disposition or to role/function, it can be ascribed to deliberate choice.

D *'Ignorance' as a deliberate choice*

In some cases scientific knowledge is bracketed, ignored, jettisoned or avoided because it is essentially peripheral to, or may even obscure, the real issue. Here, 'ignorance' is constructed as a *deliberate choice*.

In one panel interview, the respondent, DM, characterised scientists as 'people who have an incredible faith in science'. When subsequently asked whether he thought radiation can have any good effects, he remarked, 'There must be one or two, but I can't be bothered thinking about them.' This self-imposed limitation on knowledge-seeking also emerges in extract (F). Just prior to this passage, the conversation has been about the various types of radiation and subatomic particles. Then the interviewer asks how X-rays fit into the discussion. The interviewee replies:

> BW: I'm not sure . . . I know that the net effect of the whole damn lot is that if you get an overdose it's curtains and I don't really need to know any more.
>
> *(Panel: BW. 1.335)*

In both these cases there is a reason for not knowing more: extra knowledge is redundant, or a distraction from the central issue – that radiation is dangerous. It seems that knowledge of the processes and uses of ionising radiation is represented as of marginal relevance. The 'ignorance' or deficit is no longer a state, but a positive choice. We have seen that the state of 'ignorance', signalled by the lack of a scientific mind, suggests subordination to science as a good or perhaps unavoidable authority. We have also noted that 'ignorance', as a corollary of the division of labour, implies a social and practical functionality and a collaborative relationship with science. In the present instance the apparent choice to curtail scientific knowledge is part of an effort to maintain social independence from science and, possibly, to challenge the authority of interests using 'science'. This is noteworthy because critique is conducted by changing the register of the debate: it is no longer a question of working through scientific arguments to find a different truth, rather, science is irrelevant – rendered peripheral to, indeed occluding, the real issues at stake. This contrasts with Smithson's view that 'intended ignorance usually performs defensive social functions'.[31] While this can undoubtedly be the case, in the present instance we have a rhetoric of intentional 'ignorance' being mobilised to challenge or attack the relevance of a given body of expert knowledge to the 'real' issue at stake as perceived by the speaker.

This discourse of ignorance also contrasts with a more obvious antagonistic posture toward science which directly challenges the interests behind a given type of scientific knowledge. For example, one of our Radon Survey Volunteer sample applied this type of discourse to the issue of why Radon had recently become so interesting:

(G)

> JM: I don't really know why they're having such a push on a national survey. It's the fashion to do surveys on radiation because it's a popular issue. People like yourself and people in science can go and get doctorates . . . you'll certainly never cure background radiation. it's just job creation . . .
>
> *(Radon: JM. 78)*

Here, the scepticism reflects this speaker's status as an ex-physics graduate who confidently believes that Radon is essentially a non–issue. The self-ascribed expert knowledge about the status of Radon both in itself and relative to what he considers to be more urgent problems such as overhead power lines, allows him to be sceptical about the Radon survey. In seeking out reasons for it, he settles, perhaps ironically, upon job creation. Our second example shows the speaker formulate the Radon issue as a deliberate distraction from the central issue of the danger of the nuclear industry:

(H)

I: What do you think of as natural sources [of radiation]?

PD: The sun . . . em . . . you which is . . . em you know . . . which is dangerous but we're living on this planet and er have our ionospheres and our stratospheres and all these things which appear to protect us so we can get the benefits of it without the dangers of it. Well one of the natural radiations which . . . radioactive substances which the industry's jumped on recently is this Radon gas which is naturally occurring so that they can divert people's attention from them so that they can say 'Oh look, well you know, we will be able to seal . . . [nuclear waste products].

(Panel: PD. 1.372)

Here, it is advanced that the Radon survey is not simply a survey – but has specific political spin-offs for the nuclear industry. That is, he has placed the Radon issue in a larger context – it has become a resource for the nuclear industry and as such is to be condemned. Indeed, it is almost as if Radon as an issue has been constructed by bad science (science with suspect interests). Against this view, the speaker deploying an ignorance-as-choice discourse was questioning the usefulness of knowing scientific detail for broader political questions. The texture of science, the minutiae, the arcane – all these only obscure the real (political) issues. Here, to assimilate specific scientific knowledge (for example, about ionising radiation) would be to assimilate a particular (inferred) set of social interests and values which are anathema (as opposed to merely beyond one's capacity or role).

Concluding remarks

Table 5.1 summarises the three types of discourse that 'package' absence. In each case we have picked out the projected social relations between expert and speaker. The use of particular discourses of ignorance to account for manifest absence of knowledge sets up a relation to science in which the latter comes to be contrastively constructed.

As I have already noted, the last column entails a particular reading of the discourses of ignorance that situates them as resources used to fashion a relation to science. However, as the differences with Smithson's analysis of the use of volitional 'ignorance' throw into relief,[32] there is no invariant connection between a discourse and its function. These types of discourse of ignorance can be used in very different ways to evoke rather different relations to science. For example, discourses that

Table 5.1 *A typology of discourses of ignorance*

Discursive form	Status	Lay/science relation
Non-scientific mind	Mental constitution	Subservience/dependence
Not my job	Division of labour	Coexistence/co-operation
Not interested/relevant	Deliberate choice	Moral/political challenge

construct 'ignorance' in terms of division of labour need not necessarily imply peaceful coexistence or co-operation between the different roles. We see this in the ways that Cumbrian sheep-farmers – as discussed in Chapter One – construed their role in relation to the scientists of the Ministry of Agriculture, Fisheries and Food.[33] As we have seen above, the Sellafield electricians could identify a specific overarching goal that bound them to the scientists and health physicist (i.e. the smooth running of the plant), and this overarching goal could be served by 'ignorance' of irrelevant knowledge. By comparison, the sheep-farmers saw their relation with MAFF scientists as structured by goals that were much more ambiguous. On the one hand, they were clearly aware that both they and the scientists shared the common goal of monitoring and reducing radiation levels. On the other hand, they perceived that the scientists were systematically devaluing the farmers' local knowledge; indeed, the scientists were seen as pursuing the factional goal of protecting their own expertise and superiority. Here, the farmers' local knowledge was being constructed by the scientists as 'ignorance' for purposes of status, power, or whatever. The division of labour discourse comes to take a more cynical cast in which antagonistic power relations come to be identified.

To reiterate the general point, discourses of ignorance can each play a variety of roles in the construction of a social relationship to science. The types of discourse of ignorance that I have described reflect the means by which people can tacitly reflect on, and articulate, their social relationship to science and its institutions. This chapter has attempted to put such discourses upon the agenda of research into the public understanding of science and technology. To raise this issue is nevertheless to remain circumspect about the centrality of such reflection. We do not claim that the reflections (upon ignorance and absence) that we have documented are regular events in people's daily lives. However, we do contend that they are resources: people can, when necessary, reflect upon the status of their knowledge and their relations to science for a variety of purposes.

Moreover, the resources upon which people draw reflect and mediate their broader social identities: the sheep-farmers, the electricians, the secretary all have at their disposal a variety of representations of themselves as particular sorts of persons. We have suggested that the encounter with scientific knowledge, in this instance occasioned by our interviews, provides a setting in which people perforce reflect on the epistemological status of their knowledge. In the process, our participants draw upon their social identities to resource their understanding of their own understandings, and of their relation to scientific knowledge and its institutions. However, I suggest that this is not a simple one-way process. Science is not, even in this context, simply an opportunity for a rehearsal of self: it also serves as an 'Other' against which these social identities can be delineated – that is, both 'worked out' and 'worked at'. Further, given the ambiguous character of science – at once monolithic and non-unitary, radically 'Other' and incorporated into the modernist self,[34] science in its various forms can begin to impact upon and potentially affect those social identities. To get at these dynamics of rehearsal, delineation, and development of social identity in relation to science would require more textured, immersed, and longitudinal study.

In sum, in contrast to those research efforts – 'knowledge description approaches', as I have called them – which more often than not end up as mere catalogues of stereotypes, errors, or deficits, the present analysis, within the limits specified above, has explored how the social construction of science proceeds through the contrast to self (for example, one's own 'ignorance'). In the process, what emerges is the implicit discursive formulation of the relations of power between self and science. When people reflect upon their ignorance of (scientific) knowledge, they reveal their knowledge of ignorance and its social uses.

NOTES

1. 'Ignorance' is placed within inverted commas because I would like to bracket any negative or denigratory connotations that it carries. I do not use inverted commas for the phrase discourses of ignorance because, as I argue, these are used by participants precisely to construct the moral or political status of their 'ignorance'.

2. Bodmer, W., *The Public Understanding of Science* (London: Royal Society, 1985).

3. Ibid., 12.

4. Beveridge, A. A. and Rudell, F., 'An evaluation of "public attitudes toward science and technology" in science indicators: the 1985 report', *Public Opinion Quarterly*, 52 (1988), 374–85.

5. Durant, J. R., Evans, G. A. and Thomas, G. P., 'The public understanding of science', *Nature*, 340, (6 July 1989), 11–14. Also see Evans, G. and Durant, J., 'The understanding of science in Britain and the USA', *British Social Attitudes: special international report*, edited by R. Jowell, S. Witherspoon, and L. Brook (Aldershot: Gower/Social and Community Planning Research).

6. Wynne, B., 'Knowledge, interests and utility', paper presented at The Science Policy Support Group Workshop, Lancaster University, 1988.

7. The continued currency of this approach in academic circles should not be doubted. See, for example, Bodmer, W. and Wilkins, J., 'Research to improve the public understanding programmes', *Public Understanding of Science* 1 (1992) 7–10; and Miller, J. D., 'Towards a

scientific understanding of the public understanding of science and technology', *Public Understanding of Science*, 1 (1992) 23–6.

8. For example, S. Moscovici, 'On social representations', *Social Cognition*, edited by J. P. Forgas (London: Academic Press, 1981); Moscovici, S., 'The phenomenon of social representations', *Social Representations*, edited by R. M. Farr and S. Moscovici (Cambridge University Press, 1984).

9. Markova, I. and Wilkie, P., 'Representations, concepts and social change: the phenomenon of AIDS', *Journal for the Theory of Social Behaviour* 17 (1987) 389–409.

10. Herzlich, C., *Health and Illness: a social psychological analysis* (London: Academic Press, 1973).

11. Roiser, M., 'Commonsense, science and public opinion', *Journal for the Theory of Social Behaviour* 17 (1987) 411–32.

12. Galli, I. and Nigro, G., 'The social representation of radioactivity among Italian children', *Social Science Information* 26 (1987) 535–49.

13. McCloskey, M., 'Naive theories of motion', *Mental Model*, edited by D. Gentner and A. L. Stevens (Hillsdale, N. J.: Lawrence Erlbaum, 1983).

14. Gentner D. and Gentner, D. R., 'Flowing waters or teeming crowds. Mental models of electricity', in Gentner and Stevens, ibid.

15. Kempton W., 'Two theories of home heat control', *Cultural Models in Language and Thought*, edited by D. Holland and N. Quinn (Cambridge University Press, 1987).

16. Collins A. and Gentner D., 'How people construct mental models', *Cultural Models in Language and Thought*, edited by D. Holland and N. Quinn (Cambridge University Press, 1987)

17. Quinn N. and Holland, D., 'Cultural models in language and thought: an introduction', in Holland and Quinn, ibid., note 15.

18. Quinn, N., 'Convergent evidence for a cultural model of American marriage', in Holland and Quinn, ibid., note 15.

19. Holland, D. and Skinner, D., 'Prestige and intimacy: the cultural models behind American talk about gender types', in Holland and Quinn, ibid., note 15.

20. D'Andrade, R., 'A folk model of the mind', in Holland and Quinn, ibid., note 15.

21. Ravetz, J. R. 'Uncertainty, ignorance and policy', *Science for Public Policy*, edited by H. Brooks and C. L. Cooper (Oxford: Pergamon, 1987).

22. Ibid., 82.

23. Smithson, M., 'Toward a social theory of ignorance', *Journal for the Theory of Social Behaviour* 15 (1985) 152–72. Quote, p. 154.

24. Ibid., 154.

25. For example, Potter, J. and Wetherell, M., *Discourse and Social Psychology* (London: Sage, 1987); Gilbert G. N. and Mulkay, M., *Opening Pandora's Box: A Sociological Analysis of Scientists' Discourse* (Cambridge University Press, 1984).

26. Wetherell, M. and Potter, J., 'Narrative characters and accounting for violence', *Texts of Identity*, edited by J. Shotter and K. J. Gergen (London: Sage, 1989).

27. The function of discourse in the broader ideological or institutional context has been investigated by numerous authors. For instance, the impact of more distant contexts, such as institutional fiscal constraint, upon discourse has been considered in education. For example, see Mehan, H., 'Language and power in organisational process', *Discourse Processes* 10, 291–301. Such studies demonstrate how talk addresses the immediate face-to-face circumstances and the specific conditions of distal contexts. I will not be examining the

sociological status of these contexts. This issue has been brilliantly crystallised in Knorr-Cetina, K., 'The micro-social order: towards a reconception', *Action and Structure: research methods and social theory*, edited by N. G. Fielding (London: Sage, 1988). In the present context, I will treat talk about 'ignorance' as if it is being conversationally directed to the distal audiences that comprise science.

28. As a methodological tool the notion of the ideal type has the advantage of allowing the analyst to abstract from otherwise unruly data some sort of essence. Any such abstraction is necessarily hedged with the proviso that it is a partial fabrication. None the less, in the present circumstances, the value of the ideal types as a heuristic is taken to outweigh their status as fictions. For the classic formulation of the ideal type, see Weber, M., *The Methodology of the Social Sciences* (New York: Free Press, 1949). An excellent critique of the ideal type is provided in Parkin, F., *Max Weber* (Chichester: Ellis Harwood, 1982).

29. Halliday, M. A. K., *An Introduction to Functional Grammar* (London: Edward Arnold, 1985).

30. For a more detailed account of the division of labour between science and self, see Michael, M., 'Lay discourses of science: science-in-general, science-in-particular and self', *Science, Technology and Human Values* 17 (1992) 313–33.

31. Smithson, ibid., note 24, 156.

32. Ibid.

33. Wynne, B., Williams, P., and Williams, J., 'Cumbrian hill-farmers' views of scientific advice'. Evidence presented to The House of Commons Select Committee on Agriculture Investigating The Chernobyl Disaster and the Effects of Radioactive Fallout on the UK published in the Agricultural Select Committee Report, *Chernobyl: the Government's Response*, vol. 2: Minutes of Evidence (London: HMSO, 1988); Michael, M. and Wynne, B., 'Misunderstanding and myth: the case of radiation', paper presented at the Annual Meeting of the British Association, Sheffield, 1989.

34. For example, each of the following authors have highlighted the relation of science to social identity: Holton, G., 'How to think about the anti-science phenomenon', *Public Understanding of Science* 1 (1992) 103–28; Giddens, A., *Modernity and Self-Identity* (Cambridge: Polity, 1991); Beck, U., *Risk Society* (London: Sage, 1992). However, none examine this relation in the close empirical detail that it deserves.

6 Insiders and outsiders: identifying experts on home ground

ROSEMARY McKECHNIE

Expertise in context

The public understanding of science is situated in a changing theoretical landscape. Debates about modernity, post-modernity, and globalism are throwing into question significant conceptual categories that social science has previously taken for granted. In this flux, the relationship between science, technology, and publics has become a central concern of social theory. Western society is depicted as increasingly dependent on specialised roles and institutions that are associated with specialised knowledges and competences. Accordingly, theoretical frameworks are being developed that allocate key roles to concepts relating to perceptions of risk, dependence on expert systems, and trust.[1] Within these debates the 'local' has assumed a new significance in constituting identity-based responses. However, empirical studies of the relationship between the local and wider society are sparse. This chapter is based on ethnographic research, focusing on local interpretations of expertise relating to ionising radiation in the Isle of Man. It examines the social and cultural interpretations of science that emerge from the mundane transactions of people in 'micro-social' situations. This approach is in tune with what Knorr-Cetina calls methodological situationalism, which 'demands that descriptively adequate accounts of large-scale social phenomena be grounded in statements about actual social behaviour in concrete situations' (1988: 22).

In the following pages I describe how authoritative knowledges associated with science are assumed, attributed, and evaluated in practice, within both lay contexts and institutional settings on the Isle of Man. It is proposed that there are many similarities in the ways expertise is constituted in 'informal' and 'formal' settings, and that drawing boundaries between the two is far from straightforward. The aim is to show that the assumption or attribution of expertise is a fluid process of identification and negotiation based on the 'ground rules' of cultural settings. The Isle of

Man may be a unique 'locality', but the processes defining the fracture lines between science and the heterogeneous Manx population, which this chapter explores, echo significant themes found in other chapters of this book. According validity to lay-groups' theoretical structures, and examining how these are realised in action, together present a critical challenge to analytical boundaries drawn between publics, institutions, and science. Rather than taking as a starting-point a line that divides 'science' from the rest of society, the aim here is to put science back into its cultural and social settings, where everyone, scientist, and layperson alike, is actively partici-pating in the processes of identification that create boundaries and give substance to powerful concepts such as 'expertise'.

'The expert' has come to play an increasingly important social role. However, we know very little about the basis of the credibility of expertise.[2] Understanding how 'experts' are identified, what 'expertise' means, and how it is related to the structures that underpin society, requires close observation of 'expertise' in action, in practical situations. One aim of this chapter is to show that setting 'the public' in context is a necessary first step. The empirical focus of this chapter is not on how an issue brings science into conflict with a community, but, on how, within one definitional context, processes of identification maintaining internal and external cultural bound-aries in the face of social and political changes, have important consequences for local interpretations of expertise and authority. Concepts such as 'public', 'science', or 'expertise' are social achievements, subject to differing and competing definitions. The fluid boundaries which define and oppose 'science' and 'publics' are constantly shifting, dissolving, and reappearing. Science and publics are situated not in oppo-sition to each other in a vacuum, but in a complex of relationships.

Locality and identity: Manx and comeover

The Isle of Man provided an interesting location for research into public interpret-ations of ionising radiation. The island is something of a constitutional oddity.[3] It is also relatively self-contained institutionally. Manx political specificity is strongly defended by the governmental body, Tynwald. Issues that lead to clashes with the UK are important; each can be viewed as a negotiation of the island's status. Some are economic – transport is particularly difficult – others concern moral issues such as birching or legislation concerning homosexuality. Environmental issues provide another field in which the negotiation of control and sovereignty can take place. These included debates on how the island could be brought into line with European and UK environmental directives, but the impact of external pollution on local environments was also a worry. Practically, its location in the Irish sea, near the Sellafield reprocessing plant and Heysham nuclear power station on the English coast, meant that the issues discussed by Wynne in Chapter One were topical there, including the restrictions on sheep-farming. Unlike other areas where public responses to such issues have been studied,[4] the population was not employed by, or dependent upon, the nuclear industry. The geographical distance involved, though slight, was reinforced by important political and symbolic boundaries. There was concern on the island about radioactive pollution, particularly emissions into the Irish sea. The position adopted by Tynwald, calling on the UK government to put pressure on the Sellafield reprocessing plant to reduce emissions to zero, was modified after a popular petition was presented to Tynwald calling for debate on the issue, and increased disquiet after Chernobyl. In 1986 Tynwald called for the closure of Sellafield and was, at the time this research was carried out (during 1988), on the point of producing a document clarifying its views on the UK nuclear industry as a whole.[5]

The symbolic boundaries based on the legislative differences which set the island apart were reinforced by issues that bring the island into conflict with the UK. However, insular social space was also demarcated by boundaries which were in the process of redefinition. The island has been a tax haven since 1961. As a result the financial sector has come to dominate the island's economy. The industrial sector remains undeveloped, and tourism, of primary importance until recently, is failing. The fiscal changes have had far-reaching repercussions. The accelerating depopulation of the 1950s has been reversed, and fears have been expressed about the growth in population in terms of over-development and the inadequacy of the island's amenities. Since the influx of new residents has been controlled by work permits, most permanent new residents are either professionals or wealthy. There is some friction, partly because native Manx people have been hit by failures in the non-financial sectors of the economy and property market inflation.[6]

In local parlance, the UK is talked of as 'across', and incomers are known either as 'comeovers' or 'stopovers', depending on how long they are perceived to have lived on the island. There is not room here to go into the complexities of the incomer/insular relationship, though the distinction is important and will be touched on later. The opposition of Manx and 'comeover' is central to representations of life on the island. This stereotypical opposition 'glosses over' differences in both groups. The varied incomer population is more clearly differentiated than the Manx community. While, for the Manx, diversity makes the community,[7] new residents form a less cohesive group. People coming to the island are placed principally by their socio-economic bracket. Some who have lived on the island for a time, particularly those who work, are integrated into the local community to a greater or lesser extent. However, the question of being Manx or not arises still, and, as I will touch on later, is highlighted by any discussion of local issues. Though I am concerned with cultural identity in this chapter, this should not be taken to mean that I am looking at the Manx as a homogeneous population, or 'real' Manx in isolation. Rather, I am concerned with the way that people living on the island are, from different viewpoints, engaged in definitional activity which creates and elaborates significant local boundaries, with profound consequences for issues that enter into the island's definitionary field.

Ethnography and science: some methodological questions

Ethnography entails taking note of much that does not immediately seem relevant, through getting to know people, and how they relate to each other and oneself. As Becher has noted, the distinguishing feature of anthropological research is that it is about experience rather than gathering data: 'for anthropologists methodology is the internal apprehension of relationships and their transformation through cultural meaning' (1989: 39). Interview situations are artificial. It might be contended that this is particularly true when the subject of the interview is science. As Michael notes in Chapter Five, respondents are not just talking to the interviewer, but are in a situation in which they are addressing science. What they *do not* say in their presentation is an important facet of how they work with ideas about science and their relationship to science. Denial of knowledge and refusal to participate by answering, can be positive strategies to protect personal identity. This ethnography was conceived of as complementary to the discourse approach of Michael. The common starting-point is not individuals' lack of knowledge about scientific 'facts' or processes, but how people reflect on the status of their own knowledge and situate themselves *vis à vis* science and *vis à vis* others in relation to science. Unravelling the way self-definition frames a relationship between individuals, groups and 'science' uncovers how 'ignorance' rests on not only structures of difference, but also relations of dependence, co-operation, or challenge: relations of power. The ethnographic material used is similar to the interviews analysed by Michael in many respects, but, rather than focusing on individual constructions, ethnography locates accounts in the context of the multiplicity of other social relations in which people are involved.

Ethnographic methods have successfully been used in the context of the laboratory to dissect the way scientific meaning and authority are constructed.[8] Outside the laboratory, science may play only a small part in people's lives. Their perceptions of science cannot be divorced from their perceptions of the complex web of social and institutional relations in which it is embedded. Studying lay interpretations of science in their own terms decentres science. In a context where the particular domain of science under scrutiny, ionising radiation, evokes almost universal denial of knowledge, there is often nothing but 'the social' to study. As Callon has pointed out, the interpretations of sociologists of science have been marked by a curious asymmetry, their willingness to accord equal validity to all actors views of nature and scientific/technical theory has not extended to actors views about society:

> [they] act as though this agnosticism towards natural science and technology were not
> applicable towards society as well. For them nature is uncertain, but Society is not . . .
> Both the identity and the respective importance of actors are at issue in the development
> of controversies, and ignoring the fact that identities of actors are problematic risks
> badly distorting the situation. (1986, 197)

The approach taken here has much in common with what Callon has termed 'a sociology of translation'.[9] However, anthropologists have noted that the perennial problem of 'translation' from one cultural system of meaning to another is no less pressing within the confines of one's own language. Here, the researcher shares assumptions with those being studied, and, like them, takes for granted the meaning of significant categories.[10] A further point arises in relation to the aim of giving an unweighted account of different actors' views of the world.[11] While it is desirable to dislodge science and social science from a privileged position, this aim should not sidestep efforts to place interpretations within a framework that recognises structures of power that exist independently of any discourse about them.[12] This implies the recognition of structures that both frame the context of research, and the relationship between subjects and researchers. The approach adopted here does not give precedence to scientific interpretations of events, but examines the processes that lend authority to 'expertise', focusing on how these are created and recreated in a variety of social relations. Latour (1986) urges social scientists to follow science out into the world. Here I take a different path by following science from the world, tracing its path partway back. The assumption is that science, and its practitioners, are not the only, and perhaps not the principal, actors involved in the social construction of scientific authority. It is not a one-way process. The whole of society participates in identifying 'science' and 'expertise', as it does in the identification of any important symbolic boundary.[13]

Local knowledges: identifying local expertise

When describing the aims of our research to people on the island, I left the parameters of definition as wide as possible. Amongst other things I was interested in finding out what ionising radiation meant to people (if anything) in local terms. It was almost universally assumed that my research must be concerned with radioactive pollution, and this was further condensed to relate to two main issues: Sellafield's discharges in the Irish sea and Chernobyl fall-out. These were recognised as 'important' issues for the island, and almost everyone could refer to items of news in the local media.[14] Any further probing usually resulted in statements of denial of knowledge. However, unlike the responses given in interviews, these denials were accompanied by suggestions about who did have knowledge about the subject. People were surprised that I should want to talk to them; what they did know, and what they assumed I wanted to know, was who could give me an authoritative account of issues relating to radiation on the island. They were, in effect, giving me access to a shared, local mapping of expertise. In the following pages I describe how individuals' relation to science were grounded in social relationships, treating this map as a schematised network of the significant relational structures through which science is interpreted. The aim is to render explicit the processes that identified 'expertise' in local terms, examining the ways in which specialised knowledges were constructed and evaluated as trustworthy and authoritative.

This analysis is mainly concerned with a network of knowledge common to the whole island, but before examining this I would first like to consider the more proximal mapping of knowledge within the village where I lived. The processes involved in identifying and evaluating those who had relevant knowledge are similar to those at work in the wider context of the island as a whole. Some of those whose views I was encouraged to solicit predictably held positions of authority within the community, such as local politicians, religious leaders, and the headmaster. Others were less obvious, for instance the local fishmonger. He was often asked whether the fish he sold were likely to be affected by the radioactive pollution of the Irish sea. His customers wanted to know if fish had been landed on the west side of the Island, or Ramsey in the east (from where Sellafield is visible). His responses knitted together his own practical knowledge about the fishing patterns of the boats, the movement of fish shoals, and Irish sea currents, with some elementary physics concerning heavy elements. His practical knowledge was respected, and his standing as a local who had steadily built up his own business enhanced the authority of his statements. He epitomised good 'common sense'.

The inhabitants of this tightly knit village shared a detailed body of common knowledge identifying each individual. The process of identification was dynamic, and what was known about others was constantly under revision. The views that people expressed could be interpreted in light of prior knowledge about them, and at the same time added to the pool of knowledge that identified them within the community. The authority to speak about any issue was allocated on the basis of a number of factors related to established local status and personal experience. This authority was evaluated in terms of several significant dimensions; general knowledge, intelligence, and presentational skills were balanced with participation in local issues and specific local knowledges. The one factor that appeared of little consequence in this context was scientific knowledge. Indeed, those villagers I had been directed towards as sources of information were themselves very modest about the limits of their scientific knowledge, though they were more forthcoming with their views about the social and political aspects of these issues on the island. They, like everyone else, advised me to speak to others who 'really knew' about the issues.

The individuals identified as 'local experts' within the mapping of knowledge that covered the whole island were a varied group. They included public figures such as politicians and government officials, as well as a few individuals who were perceived as having taken an active role in shaping Manx legislation relating to the British nuclear industry. Three MHKs (elected representatives: Members of the House of Keys) were pointed out to me by people as 'the MHK I should talk to about radiation'. One of these had entered politics through interest in environmental issues when living away from the island, and another had a scientific background and was vocal in debates about local environmental projects. The third was rather different, a local sheep-farmer himself, he had created controversy by defying post-Chernobyl restrictions and was very vocal in his opposition to Sellafield. In addition, the activist who organised the petition that was presented to Tynwald in 1984, leading to the debate that resulted in the 'close Sellafield down' policy, was well known and had acquired the nickname 'Sellafield Sue'. It was she, rather than government officials, to whom many people turned immediately for reassurance and advice when they heard that the cloud of Chernobyl fallout had passed over the island. All of these people could be seen as laying claim to specialised knowledge in the way that they had made public their views.

The popular association of these figures with the issues might give the impression that the insular population accepted their authority. Nothing could be further from the truth. Very few recommendations were given to me without some further remarks being made about the individuals concerned. These had little to do with science or radiation, but generally were of a personal nature. Many of these remarks could be categorised as 'gossip', relating to the personal lives of public figures. They concerned allegations about honesty, dependability, and sexuality. These straightforward, and sometimes damning, moral evaluations were drawn from a reservoir of 'common knowledge' which could be used to frame these figures' views on any subject. Other remarks reflected on their effectiveness on more familiar ground. The authority of one official, for example, was undermined because he had sponsored the showing of a video on farm accidents at local schools. It was said that this had been unwise since it had scared some of the children half to death. In a professional capacity failed business ventures signalled lack of judgement or competence: 'He's had several things on the go in the last few years, none of them came to anything.' Similarly: 'If he can't make a decent pizza how can you trust what he has to say about the nuclear industry?' Scientific knowledge played a relatively unimportant role in the credibility of these figures. Their status as 'expert' was judged in terms of their integrity and competence in day-to-day life.

There is a tendency within research concerning public responses to scientific issues to dismiss as irrelevant moral evaluations of persons and institutions. As Callon (1986: 198) has pointed out: 'sociologists tend to censor selectively the actors when they speak of themselves, their allies, their adversaries, or social backgrounds'. However, the powerful role that such information can play in the evaluation of an agent as someone who can be trusted and respected in important decision-making is not limited to the politics of 'face-to-face communities', but can be found in many decision-making arenas, and even in the heart of scientific institutions themselves.[15]

This 'framing' of those claiming authoritative knowledge made explicit the significant dimensions of local evaluations. One factor that came up frequently in descriptions of public figures were development projects that were the focus of local debate over the future of the tourist industry. One local group I talked to, for instance, insisted that we should go to a local hotel that was threatened by a large development project to discuss the issues relating to radiation. Several public figures who were associated with environmental issues were also vocal in the debate concerning this development. The group wanted to draw parallels with the way these figures viewed the destruction of the hotel, which was situated on a beautiful spot popular with local people, and their views about the environment. The position of one figure in particular, who had decried local opposition to Sellafield as based on irrational fears and ignorance of the science involved, was compared with his support of the new development. His commitment to 'progress' at the cost of changing the local environment was presented to me as antithetical to local values.

This description of the way 'expertise' was identified within the Manx context and embodied in social relationships shows trust and authority to be heavily contingent. The possession of scientific knowledge did not appear to carry much weight in the evaluation of expertise, but was overshadowed by perceptions of personal integrity, and effectiveness in local issues. However, scientific knowledge was not just being ignored, rather it was engaged with in an oblique way. The next section looks at how the framework that grounded interpretations of expertise firmly in local issues and relations, also provided a basis for undermining the structures that set science apart.

Common sense: drawing the lines

One of the most important factors involved in local evaluation of those perceived as possessing 'expertise' was the style in which claims to authoritative knowledge were made: the *way* views were presented, rather than the views themselves. Commenting on the MHK with a doctorate, one person said: 'Doctorate in what I wonder, I'm always suspicious of people who use titles, especially when they are not relevant.' Here a claim to authority based on external, 'objective', professional credentials was being challenged by local conventions of self-presentation. Individuals are credited with authority and respect only if their self-presentation is consistent with local values. Any sort of assertion of authority in terms of esoteric knowledge is dangerous and often leads to loss of reputation rather than the reverse.

Expressing any claim to superiority is a dangerous business on the Isle of Man. Egalitarianism – 'we're all the same here' – is common in tightly bound communities, even in the face of obvious inequalities of wealth, education, or professional status. The shared commitment to this ideal is achieved by sanctions that undermine any claims which threaten the group's value-system. The use of such sanctions was pointed out to me frequently as a typical Manx trait.[16] The way this process works within the Manx community is encapsulated by the derogatory label 'when I'. Those who have been 'away', who boast of their successes or experiences, or who dare to reflect negatively on island life are dismissed as 'When I's', shortened from the phrase, 'When I was in . . .'. Status and knowledge gained elsewhere are put in their place in local terms. Reputations are made and destroyed by common recognition of this value-system, but those who are not familiar with it, or who fail to read the signs, can remain happily oblivious to the fact that they are considered fools. In the insular Manx community plain speaking is positively valued. Presentations in difficult or esoteric language are frowned on as 'airs and graces', and the speaker dismissed as arrogant. This re-evaluation in local terms is, in effect, a reversal of the dominant (and on the face of it unchallenged) position of educated and scientific idioms. The rhetorical power of science is deflated and 'ordinary' language assumes a pre-eminent normative position within the defining bounds of local culture.

One particular set of people identified as possessing specialist knowledge in relation to radiation gained credibility in these terms. Sellafield's discharges in the Irish sea and the contamination of sheep pasture by Chernobyl fall-out were considered to affect two groups in particular: farmers and fishermen. The insular Manx population identified a few key individuals as representatives of these groups. All of them had earned their due respect over the years. They were not presented to me as having a particular view, nor were they known to have particular knowledge of the issues and science involved. It was assumed, however, that they would have a sound idea of what the issues meant in terms of farming and fishing. They did not escape completely the criticisms levelled at those assuming a position of public visibility. For instance, the parting comment of one person after recommending that I speak to a fisherman was, 'he'll talk to you, he likes the sound of his own voice that one'. However, they were seen as able to translate abstract issues into practical meaningful terms, they were trusted mediators. As representatives of local interests their ability to speak *to* people in a comprehensible way as well as *for* them was important.

The credibility that their practical expertise gave them underscored an important aspect of perceptions of theoretical knowledge, namely its dislocation from practical experience, epitomised by the cliché 'that's all very well in theory, but . . .'. Lack of familiarity with the empirical and the social context can make nonsense of abstract theory *in situ*. A subtle evaluation of one public figure's opinions on radioactive

pollution emerges from one fisherman's account. Rather than make any direct statement, he switched from talking about the MHK's public statements about radioactive
pollution's effect on the fishing stock to an incident concerning the fishing of basking
sharks. This same MHK had been horrified to discover that local fishermen were
catching basking sharks that had been appearing near the coast. The fisherman said
he realised that this might seem terrible:

> These big, plankton eating things, like whales ... but people from the city don't
> understand. When they are around like that in hundreds and there is a market for
> them, there is no harm in the locals catching a few and making some money ... the
> small boats have a hard enough time ... they knew what they were doing.

The parallel was clear, environmental opinions based on abstract theories did not
necessarily relate to the 'real world'. Further, the world of those who held such
abstract views was distanced from the reality of those whose knowledge was based
on experience. Educated people with good intentions did not necessarily know what
they were talking about. The superiority of abstract knowledge was undermined by
comparison with knowledge gained from experience.

The undermining of abstract, theoretical knowledge by 'down to earth' observation
is an everyday occurrence, not restricted to the Isle of Man, but, as Geertz points
out, there is nothing straightforward about 'common sense' and there are a number
of reasons why it should be treated as a relatively organised body of considered
thought.[17] The opposition of scientific knowledge and commonsense opposes 'us'
and 'them': it is to do with social identification. It undermines abstract thought by
presenting what we 'know' to be true in its place. The wish to present any public
as capable, reasonable thinkers can lead to accepting 'commonsense' at face value.
There is a danger that slipping into uncritical reproductions of commonsense observations, whether this is to denigrate them or to laud their wisdom, will reify the
process setting the two in opposition. Two (at least) interpretations of 'the facts' are
at issue. In order to analyse the resulting situation, it is necessary to examine both –
and also the opposition itself. Such an opposition is not 'natural', it symbolises a
drawing of lines which can lead to a serious breakdown in communication, and
pre-empt negotiation of shared meanings and knowledges.

The way 'common-sense' interpretations are curiously invisible from some viewpoints, but powerful motivating forces from others, is central to understanding perceptions of expertise. Looking at interpretations of science or technology is not a
matter of following the linear trajectory of an artefact from one definitional structure
through others, which alter it in minor ways. Rather, the observer is faced with
simultaneous, conflicting interpretations of the totality of the social and natural
world. However, while all 'worldviews' are equal in their definitional potential, dominance occurs when one structure blocks the power of actualisation of others. Here

I would draw on the work of Edwin Ardener, and his concern with 'socio–intellectual structures that regularly assign contending viewpoints to a non-real status; making them "overlooked", "muted" "invisible"', (1989: 133).[18] The 'muted' nature of some interpretations results from processes that are at work in everyday interactions ensuring that some interpretations are restricted while others carry more weight and have wide cultural validity.

The rhetorical power of science is integrated into Western cultural 'ground rules'. As noted by several authors in this book, 'ordinary' common sense appears inarticulate and incoherent in comparison with the persuasive explicative power of scientific idioms. People are aware that their views and their way of articulating things are categorised as inappropriate and of lower status in wider contexts. As Michael points out in the previous chapter, the 'deficit' view of the public has entered into the popular consciousness (see also Wynne 1991). This does not mean to say that publics accept scientific accounts passively. Lay interpretations may be muted in relation to science, but they do provide the basis for positive re-evaluations of local knowledges in their own terms. Scientific knowledges seeking uptake enter into contexts that are active, generating meaning. In so far as people's worldviews coincide with dominant structures, they can expect their definitions to be realised in action; their categorisation of the world can be synonymous with their experience. In contrast, 'muted groups' are aware that their own interpretations are viewed as inferior, and their worldview also contains a reflexive consciousness of the denigratory way they are defined within dominant structures. They are more likely then to have developed ideas about the social and cultural relationships embedded in the status of scientific knowledge. The muting of local knowledges in relation to science renders cultural dislocations in interpretation invisible much of the time. However, these processes are dynamic, embracing a changing social world that continually throws up new situations that require redefinition, jeopardising established relations.

The widespread pollution resulting from the Chernobyl accident of 1986 provided just such an unforeseen situation. Paine's research with the Saami and that of the Lancaster group with Cumbrian farmers pointed out how these groups' own specialist practical knowledge was a treasured part of their identity, and its exclusion from official responses further entrenched the boundaries that produced perceptions of the situation in terms of 'us' and 'them'. Paine expresses very clearly how the Saami came to feel that dependence on expert knowledge delegitimated their cultural identity (1987, 1992). Similarly Cumbrian farmers[19] were confronted with policies that were impracticable and alien in their formal style and planning, while their own knowledge of local habitat and of farming practices was ignored (see Chapter One). There was no basis for negotiation between the differing perspectives, as policy was based on a scientific definition of the issue, which did not recognise any other relevant point of view. As a consequence, farmers' interpretation of the situation developed in terms of ideas about the social relationships involved quite as much as 'facts' about radio–isotopes, pastures, and sheep. This process of identification reinforced the cultural boundaries setting the farmers apart, in opposition to governmental, industrial, and scientific groups.

In fact, on the island, Chernobyl fall-out gave rise to a quite different train of events. The first response of the Manx government to the uncertain magnitude of the problem was to table a meeting with farmers' representatives. The difficulties involved in responding to the information that was coming from the UK, with its contradictions and uncertainties, were shared by politicians, officials, and farmers. The focus of discussions was on those areas over which those around the table had control; the practicalities involved in enacting the restrictions that they were told were necessary. Unlike the Cumbrian case-study, Manx farmers' knowledge formed a valuable contribution to the translation of scientific knowledge into practice. Working through the problems involved in monitoring the sheep took time and stretched the workload of the officials of the Department of Agriculture. However, they were familiar with the practical constraints imposed by hill farming, and were allowed the flexibility to develop procedures as they saw fit. Those officials who carried out the monitoring had contact with farmers over an extended period and through the many changes in the expert advice.[20] They bore the brunt of mediating between science, legislators, and farmers. One man described how he found himself in a position of having to read about the science involved because farmers expected him to be able to explain why monitoring was organised as it was. He was unwillingly being put in the position of 'expert'. His strategy was to act as a pro-active broker. Rather than laying claim to the knowledge himself, he photocopied articles in the popular scientific media he found useful and gave these to farmers, working through the text with them. Any discomfort about expert opinion and the setting of limits was thus shared. There *was* resentment amongst farmers at being told how to organise their land and animals, and relations between the different parties were put under some stress. However, there was also common agreement that the reputation of Manx lamb was important to protect. Resentment arising from the uncertainty regarding 'safe limits' was exacerbated by the feeling that these limits were tied to the perceived interests of the UK government. I shall look more closely at the boundary between local and UK knowledges in a later section, but first I would like to examine further the boundaries within the Manx context.

Insiders and outsiders: identifying the margins of science

As was stressed earlier, the island encapsulates a varied population with different points of view. However, those who live on the island are tied together by the definitional processes delimiting local cultural boundaries and shaping perceptions of expertise. Within the Manx context, any statement made is open to interpretation in terms of the stereotypes of outsiders and insiders. The categorisation of different styles of communication overlapped with cultural identification. Most of the public figures associated with these issues were not native Manx, or had spent some time away, and there was more than one reference to 'another comeover who thinks they can come here and tell us how to run things'.

Any outsiders' opinion may be used to exemplify how all comeovers are pushy and think they know everything; whatever they have to say might be discounted on this basis. This is a mirror reflection of outsiders' representations of the Manx as backwards and slow. Again, this can provide a framework within which to set any Manx point of view, or a perceived lack of point of view, as the oblique references and understatement of discourse shaped by Manx rules of play go unnoticed or are misinterpreted. The imagery that opposes Manx and 'comeover' paints a picture of a traditional rural community invaded by the rat-race with its attendant values. Day-to-day contact often gives rise to experiences that strengthen rather than challenge stereotypical representations.[21] While incomers may see themselves in a backwater where little happens (often following keenly national and international news) Manx information networks are alive with information relating to insular politics, issues, and scandals.

Several of the incomers who had taken a public stance about issues relating to radiation explained their adoption of a public stance in terms of the nature of the Manx people. One felt that feelings on the island were strong but unstated. This image of the nature of the Manx people, as 'grumblers', too shy to voice their view effectively, positively valued the role of the outsider; it took an outsider to give impetus to a public voice there. Some admitted disquiet about taking on too prominent a role, in the knowledge that 'comeovers' were often resented as 'taking over' any public matter. However, they perceived there to be a shortage of local people used to addressing both the public and the authorities. There appeared to be no local framework for organising protest in an effective way. Another criticised representations of the island as 'the eternal victim of oppression', going on to give his own account of the 'Manx condition'. Here the 'passivity' of the Manx was presented in a more negative light, as a 'malaise' linked with other problems such as lack of ambition, high figures for alcoholism, suicide, and divorce.

The common thread running through both sympathetic and critical accounts of Manx nature was passivity and fatalism. Herzfeld has noted that this is a common theme in popular and academic accounts of peripheral European populations. This 'mark of otherness' fits with criticism aimed at the inability of these areas to establish proper bureaucratic organisation, discounting their cultural organisation and expression (1987: 49–53).[22] It is important to note that this exclusion of science, in both dominant definitions of the 'other' and self-definition can be linked to a variety of social parameters, for example, class, gender, or race. This pattern of definition is marked in relation to the Celtic peripheries[23] where the emphasis is very much on the 'natural' and 'wild' character of the Celt (see Chapman 1978). One of the definitional oppositions that sets the more remote peripheries of Europe apart from dominant centres is the association of the centre with science, modernity, and rational government. The idea that science and technology were not the province of the Manx, that they were in some way unsuited to dealing with such issues, was certainly present in outsiders' accounts. These ranged from describing the Manx as insular, uninterested in wider social issues, to more mystical explanation. 'There is something about the Manx character that sets them apart . . . they are too self-contained, they keep everything inside . . . People aren't interested in anything further afield than the next street, they don't think about issues like nuclear power.' One man told me he had lived on the island for twenty years and could count on the fingers of one hand the Manx he could discuss 'issues like that' with. Another likened the Manx approach to radioactivity to superstition, saying they saw it as something like the evil eye, invisible but deadly.

This stereotypical theme was present in an ambiguous joking way in many of the asides by which native Manx responded to my questions. People followed their own denial of knowledge with an assertion to the effect that I would not find many other people there interested in that sort of thing. Allusions were made to inbreeding and emigration of the brightest. Individual assertions of ignorance in terms of being unfit to know, establishing science as distant and unknowable, were extended to the community; creating solidarity in their position *vis à vis* science. Paradoxically, however, this apparent espousal of self-disempowerment disguised how strong the power to resist external, scientific interpretations of issues was on the island. Within the Manx community there is a strong prevailing cultural basis for the re-evaluation or even rejection of scientific expertise that jars with local style and practical knowledge. Far from being a passive public, waiting to be persuaded or educated, people were actively interpreting from their own point of view and with their own aims in mind. The opposition of scientific and lay knowledges overlapped with significant symbolic boundaries between the island and the outside world, and between different groups on the island. These resulted in competing definitions of expertise on the island. Outsiders mobilising extraneous knowledge on behalf of the Manx public as a whole, ran the risk of being evaluated in negative terms. According authority to scientific expertise on its own terms threatened local structures of power, which disadvantaged outsiders who did not understand the 'ground rules' or possess relevant local knowledges.

The apparently ineffectual 'grumbling' that some outsiders perceived was actually the tip of an iceberg of very effective information sharing. The sanctions imposed by 'loss of face' through informal channels within the community was an efficient system for making sure that local opinions were well represented. The pressure exerted on representatives could be very effective. As one MHK put it: 'people are more interested in the potholes in the road but if things go wrong, then they ask questions all right, then they'll ask what we've been doing about it all the time so the stance taken is critical, that's why we did the report'. The report in question was the document that outlined Tynwald's stance on UK nuclear installations, particularly Sellafield. In the next section I follow expert and local knowledges into the institution of Tynwald.

Tracing expertise through institutions

Tynwald occupies an important position both as the main seat of responsibility and decision-making, and in its mediating position between the island and the outside world. The House of Keys, the body of elected representatives (there are 24 MHKs) and the government's 'civil service' are all housed in Atholl Street, Douglas. Each elected member is assigned a position in one, or more, of the departments that are loosely based on the UK governmental system.[24] Each department (there are 9 in all) has a minister, several elected representatives, and the equivalent of civil servants. The latter both provide advice and put into action directives from the government. Since all are gathered together within one relatively contained area, around a communal café and library, the boundaries separating each department are much less pronounced than in the larger institutional setting they emulate. The informal aspects of the system are recognised and valued, the tension between informal and formal accepted. There are some indications, however, that the balance is changing and that new circumstances, particularly related to environmental issues, are placing some strain on the institution.

In the micro-context of island politics, where the 'social distance' between the public and their representatives is very small, this negotiation is based primarily on shared local knowledge. The Manx see their government as composed of individuals with individual responsibility, competence, and moral integrity. One important aspect of representations of the Manx character is that the Manx are uninterested in politics. Local politics are said to be unlike politics 'across'. Most MHKs stand and are elected as independents. They are not expected to have a coherent policy (though, recently, new members have produced documents describing their aims and views), but are elected on the basis of personal qualities. Until recently, most MHKs have been local businessmen or professionals. Since most MHKs work within the community, and have been part of it for some time, they are visible, familiar, and approachable.[25] They are held personally accountable to their constituency for their decisions. Since personal knowledge of those in power is extensive, and there is an absence of party political rationale for their decision-making, much of the public's theorising about their actions and decisions is focused on the personal level. Personal reputation is of vital importance in this political culture.

Despite the knowledge that people had about members of the House of Keys, in many ways Tynwald was a 'black box' to members of the Manx public. They had little or no knowledge of the relevant expert knowledges used as a policy resource within the institution. Some had heard of the public analysts' office, few knew of the civil servants who were working on issues concerning radioactivity. Perceptions of what constituted relevant expert knowledge within Tynwald were very different from views expressed outside. The MHKs I had been directed to by members of the public were modest about their own grasp of the science involved in issues relating to ionising radiation, and pointed out the few individuals who they considered really did know about radiation within the institution. Again, 'real' expertise was always at a remove.

A theme which ran through all accounts was the relative lack of scientific expertise within the island. This has certainly made itself felt in the last ten years as the island has had to deal with an increasing number of environmental problems. As one MHK put it 'In a way we all have to be experts in everything here.' Not only do decision-makers have to deal with several areas at once, the government officials who are advising and later implementing regulations are relatively few in number. While Whitehall staffs departments with qualified specialists, in Tynwald one person may be covering several areas without considering themself to be 'really expert' in any of them. The few who do have a scientific background, notably in the analyst's laboratory, are called upon as 'experts' in a wide range of scientific, environmental, and technological issues. An important point emerging from consideration of both the public and governmental domains on the island, is that the 'floating' attribution of expertise contingent on knowledge, personal qualities, and experience remarked on in everyday conversations, can be reified through time and circumstance.

Over the years a few individuals have assumed or been credited with 'temporary expertise', because they possessed knowledge that was perceived as more or less related to scientific knowledge about radiation. In the absence of more specialised knowledge coming into the local context, this sort of temporary ascription has become permanent. As the compiler of the reports on which Manx policy concerning the UK nuclear industry is based explained, basically there was no one with any idea about the subject, and, as he was the person who dealt with anything labelled environmental, he had to do it. He had to work very hard to try and understand the issues, spending an inordinate amount of time over the last ten years acquiring competence in relevant scientific and technological fields. His evaluation of his work was pragmatic; even if he did not have the grounding necessary to understand all the ramifications of each issue, within the existing limitations it was as good as it could be. His endeavours were appreciated by colleagues and MHKs, the report was perceived as forming a practical basis for policy-making and presenting an informed political stance to the outside world.

The public analyst's office provided another location of 'expertise'. The incumbent analyst at the time of the 1957 Sellafield fire, had conducted a few experiments, for his own satisfaction as much as anything else, measuring beta radiation in various foodstuffs. He was then called on, in his official capacity, to measure the effects of the fire on the island. From that point on, dealing with issues concerning radioactivity were assumed to be part of the office's remit, as was presenting findings to the public via the local media. Recently, a scientist had been employed specifically to monitor radiation on the island. Though his more sophisticated ideas about how to present figures to the public (something that had been included in his training, indicating the development of 'public relations' science) made the writing of press releases easier, he was still uncomfortable with this responsibility. A scientist who had worked in a large laboratory refining techniques for measuring one radio-nuclide, he found that on the island his brief was much wider. To begin with he had to set up his own measuring equipment, even to collect his own samples (in his own car) with few resources. No one else was aware of the differences in technique that alpha, beta, and gamma-radiation required, or the sensitive standards of calibration now necessary for professional credibility.

There were marked discrepancies in actors' own appreciation of the status accorded to their 'expert' knowledge. As one man put it, that was in a way what he liked about working on the island: 'you never know what's going to happen next'. It was noticeable that others were less relaxed with the authority that their standing on the island carried. For instance, the new community physician, who found he was in the position of having to reassure people about Chernobyl fall-out without being absolutely clear on the dangers it represented. There was no 'higher expertise' to refer people to. Most of these actors had come from institutions elsewhere, there was no coincidence of 'habitus' and 'habit' for them, they were not 'at home' in their position (Bourdieu 1981, 308).[26] Their own practical experience and training had not prepared them for this position of being the sole authority for a wide variety of issues. In contrast, officials whose primary training took place on the island were relatively relaxed with the idea that they should adopt a pragmatic approach which drew on areas of expertise in an *ad hoc* way to deal with problems as they arose.

Dealing directly with the local public presented problems to those bringing expertise from the outside world to the Manx context. They were transplanted from a position in a wide scientific establishment where their particular role was defined within a network which spread knowledge and responsibility over a wide institutional area, to a position where they represented the scientific establishment *per se*. As a result, aspects of science that in their 'native world' could be taken for granted or left to others to deal with, were 'twisted off' into this new definitionary context in a disconcerting way, leading them to reflect on areas they might not otherwise have perceived as problematic. One scientist reflected ruefully on what appeared to him to be the arbitrary interest of the media. The levels of radiation monitored after the Sellafield fire, for example, provoked little interest, while recent results, which he described as minor by comparison, were interpreted as indicating the public were at risk. 'Facts and figures' were used to substantiate shifting and conflicting views. Another took the position that there had to be a safe limit, based on parallels with an area of chemistry he was familiar with. He was prepared to trust the 'real' experts on ionising radiation to define these, but was confounded when he found that the 'experts' kept changing the guidelines. He worried about how this uncertainty could be presented to the public in a way that would not undermine the scientific establishment.

The scientists could be called upon to represent the island to the outside world, notably to provide data in support of the critical stance adopted towards the British nuclear industry. They were caught in something of a double-bind, as the values of the scientific establishment, which they still considered themselves to be working within, were potentially threatened by local challenges that they were participating in. They were also aware that their expertise was vulnerable to evaluation in terms of different professional standards outside the island. They found themselves embodying a stressful position which straddled significant boundaries and was subject to pressure from both sides. The difficulty experienced by those who had to play an expert role both locally, and in wider arenas, renders explicit inbuilt assumptions current in dominant institutions about the nature of scientific knowledge as established, above political, social, and moral interests; a hierarchy of expertise based on credentials and specialisation. Like other local 'experts' already discussed, those who had come to the island with specialist knowledge were anxious to put this knowledge in perspective, comparing it with 'real expertise' which was again located elsewhere. There was varying dissonance between their ascribed status on the island and their own ideas of expertise shaped by their different professional cultures.

The above description may give the impression that the institutional identification of need and the actual 'bricolage' of available knowledge owed more to serendipity than design. It should be stressed that this is not particular to the Isle of Man, but is a much more general process. Bourdieu describes such situations as 'grey areas' of social space (1981: 310) and points out that a multitude of factors are involved:

> The institutionalisation of 'spontaneous' divisions that occurs little by little, under the pressure of events, through the positive or negative sanctions the social order exerts on organisations . . . leads to what can eventually be seen as a new division of the work of domination . . . Thus the social world comes to be peopled with institutions which no

one designed or wanted . . . [These are] . . . Not just production of bureaucracy nor of individual transactions but both the social conditions of the production of agents (inside and outside the institution) and the institutional conditions in which they perform their functions.

(Ibid., 312)

The identification of expertise within the institution was very much an extension of the local cultural context. At the same time, external professional, political and economic pressures also exerted considerable pressure.

Crossing boundaries: local knowledge and the outside world

We have returned by a circuitous route to expertise at the nexus of scientific establishments, policies, and publics, specifically to the island's response to scientific expertise concerning radiation. In contrast to the confident views they held on local issues, many people were uncertain about their own views concerning issues relating to ionising radiation. One aspect of the issue was clear – people talking about attitudes to nuclear power on the island constantly came back to an explanation in terms of interest. It was commonly said that if a proposal to link into the national grid came to pass, and electricity went down by 2p a unit, then all criticism of Sellafield would disappear. The formula that 'the island got all the problems with none of the benefits' was a phrase particularly used by politicians. However, although this provided a simple political formula, it certainly did not fully cover insular perceptions of the issues.

I came across very few people who were particularly worried about other pollution or environmental issues on the island.[27] This is partly due to the image of the island as a 'rural idyll', a haven of peace far from the technology of the modern world. This vision does, however, throw into relief any incursion of technology and modernity, so that radioactive pollution, from Chernobyl or Sellafield, is much more shocking in this habitat, envisaged as traditional and 'natural'. It is 'out of place' – a concrete symbol of the conceptual incursion of 'rat-race' society. Zonabend (1989; 9) describes the nuclear installations set in Cap de la Hague in Brittany in similar terms, as the juxtaposition of two different worlds, discordant universes, one frozen in the past the other turned to the future. The quality of Manx lamb and shellfish are a matter of insular pride. The fact that these were threatened, and thus by association the farmers and fishers regarded as the last representatives of 'real' Manx life, struck at the heart of the romantic imagery setting the island apart. Fears of pollution represent an extra dimension of a constantly encroaching violent and impersonal modernity.

It was often pointed out to me that the chimneys of Sellafield were visible from the eastern coast of the island, but the significance of this was seldom put into words. People did express fear and uncertainty about radiation, but this was not tied to any concrete ideas of how radiation acted. The mother who burnt children's clothes which were outside drying when Chernobyl fall-out passed over the island was not sure why she had done this. There was an area of silence, of uncertainty around radiation. Sometimes when I prompted responses, conversations then tended to the abstract, about science in general. The points of view expressed tended to extremes of evaluation, mobilising rhetorics of progress and danger. Scientific controversy was often brought to play in the latter; 'they can never agree' or 'they change their minds . . . in fifty years time they will decide that they've all got it wrong, and then it will be too late'. People also gave examples of human error, of experts getting it wrong, usually from the medical context where many had direct experience. However, ionising radiation itself was an unknown, profoundly so, in that it was not only invisible and impossible to perceive using human senses, but knowledge that enabled its measurement and evaluation was outwith the reach of local networks. The majority of people I spoke to expressed uncertainty, they felt that they should know more themselves, before making decisions. Further, they were doubtful of the extent of their representatives' knowledge. 'Real' scientific expertise was very much perceived as belonging elsewhere, but it was not trustworthy for other reasons.

There was a conflation of all related bodies of expertise (MAFF, NRPB, NII) with the British government; they were identified as 'UK', as representing the interests of the UK, and therefore untrustworthy. This was true at all levels; for example, a fisherman recounted the tale of how a trawler was dragged along by a submarine, only to have the incident denied by the authorities. One MHK I spoke to stated bluntly that what you get from experts is the view of the wider system they are part of:

> whether it's Greenpeace or BNFL there just doesn't seem to be anyone in between . . . As for the NRPB, they're seen by most as part of the government-cum-BNFL set, I don't know how far that is true, but they are perceived as the same and so they just don't have the credibility. Likewise MAFF are easily dismissed as just another part of the U.K. establishment.

Another MHK echoed this view: 'Sellafield is seen as synonymous with the UK government on the island, as is the NRPB . . . there really is nowhere to turn to for independent advice.' As Zonabend (1989: 177) notes, Chernobyl gave both a language and conceptual space for fixing nuclear disaster. One of the first consequences for the island was the highlighting of the local lack of expertise and technological capacity for evaluation of the situation. Following from this came the exacerbation of the distrust already embedded in the dependence on UK agencies, particularly Sellafield, which was, of course, the nearest monitoring facility to the island. The island has since acquired its own monitoring and testing equipment (some funding was provided by Sellafield for this) and asked to be included in both RIMNET (a monitoring system set up to measure background gamma-radiation), and LARRMACC (a local authority monitoring system which has established links with the Irish Government).

Conclusions

The dilemna for local 'experts' and representatives described above is that they have to identify in some way with external agencies of specialist information, yet they must at the same time identify with local idioms and structures that, to an extent, define themselves in opposition to those external agencies. Managing this ambiguity has arguably been made more difficult by the lack of pluralism in UK political institutions. Local politicians were being put in a very difficult position. One of the problems was that they had to be seen to be doing something for the island, but because they were being effectively ignored by the UK their efforts often went virtually unnoticed. The international dimension of environmental decision-making bodies, the changing configuration at both local and national level within Europe, have thrown the power relations between different units of government into sharp relief. At a recent international conference concerning, amongst other things, radioactive pollution of the Irish sea, the UK had insisted on speaking for them, despite, or because of, Manx opposition to their stance. The island threw their lot in with other 'peripheral' countries – Ireland and Iceland – establishing a contact that is increasingly well established. These events will have far-reaching repercussions for the definition and realisation of regional relations: boundaries are being redrawn.

The erosion of pluralistic centres of political legitimacy exacerbates these degenerative tendencies, leaving little in the way of mediating institutions to mitigate and inform state powers. The local dimensions of public interpretations of science and its practitioners have received too little attention, despite the fact that, in terms of nuclear power, oppositions to scientific authority have been organised at a local level from the very beginning, long before the advent of national opposition groups (see Welsh: 1993). The propensity to focus on the latter, with their technocratic discourses, takes attention away from important processes that affect the uptake and credibility of science. This occlusion of the political dimension of the opposition of interests at national and regional or local levels (ibid.), is echoed by the aims and methodology of much research that has investigated 'the public understanding of science'. This has concentrated on *individual* lack of knowledge, and inability to participate in democratic processes concerning technological decisions. This simplistic division between 'universal' centralist modern institutions and atomised individuals renders invisible the significant boundaries and relations that are forged and renegotiated around issues.

Constructions of expertise on the island have until recently been based on face-to-face politics where the assumption of authority is constantly challenged on the basis of an informal egalitarian ethos, on moral terms, or on 'common-sense' grounds. However, there is the paradox that this is at odds with the need for 'real expertise' at a local level to protect the island's interests and independence in the wider context. This only exposes the need to build bridges from this local context to increasingly cosmopolitan arenas of environmental decision-making in which small regional entities representing local interests are struggling to participate. Protecting the island's and other local interests and identities in the developing international environmental policy context will exacerbate tensions over identification, trust, and credibility which the fieldwork reported here has identified. However these tensions are handled in future practice, an important research conclusion is that scientific expertise is identified, experienced, and responded to in terms of the institutional, social, and cultural dimensions in which it is always and inevitably embedded, whatever specific form this embedding takes. Scientific and governmental institutions rarely recognise these dimensions as the bedrock of legitimacy for scientific and technological projects, hence of public uptake and understanding.

Ardener (1985) notes in his essay on the passing of modernism that the debate about rationality and relativism is a recurring one; it arises in periods of theoretical and moral shift, when the dominant category system is perceived as inadequate and is opened out to encompass hitherto muted interpretations of the world. The attention that has been focused on different publics' interpretations of science offers the possibility of relocating 'science' and 'the public' in a new framework, setting science in context and developing the terms of the debate constructively in a self-critical way.

NOTES

I would like to thank all those on the Isle of Man who found time to talk to me, 'show me the ropes' and make my stays there congenial. Also my colleagues at CSSSP Lancaster, the editors of this collection, and Ian Welsh for their help with earlier drafts of this chapter.

1. Giddens (1991) and Beck (1992) both place relations of trust and risk at the heart of their attempts to theorise modernity. Risk/trust relations become a central theme around which individual and collective identities are forged and organised, redrawing relationships between localities and dominant institutions. Douglas and Wildavsky (1982) while taking a different approach also isolate these key factors as crucial to understanding modern social forms. Douglas' grid/group approach focuses on the links between cosmology and social form. Like Downey (1988: 259) the approach taken here sees the links between ideologies and social forms as much more fluid and flexible within actors' empirical relationships and identities.

2. As Barnes and Edge note: 'the credibility of expertise cannot be established by strictly logical arguments ... judgements of credibility and evaluations of expertise are invariably and essentially modulated by the contingencies of the settings in which they are made ... We need concentrated empirical study ... to build up an understanding, not only of specific judgements and evaluations, but of the basis of credibility generally' (1982: 237).

For an evaluation of contextual studies of expertise see the introduction to Wynne and Smith (1989).

3. As a crown dependency, theoretically there is no sphere in which the UK government cannot legislate for the island, but, in practice, the insular governing body Tynwald has jurisdiction over internal matters, while the UK government has retained decision-making powers over matters of international dimension. The island's legislation is heavily influenced by Britain, but differs in several important fields, for example, labour relations, moral issues and European integration.

4. See Zonabend (1989), Wynne (1992), Chapman (1993).

5. Until this point the Manx government's policy had been to confine concern about the UK nuclear industry to statements about Sellafield, however, it was decided this stance should be reassessed following the Chernobyl disaster, and following from the Irish government's call for nuclear stations to be phased out.

6. Fears about the development of a two-tier economy, and pressure on local amenities were voiced over a decade ago (*Report of the Select Committee of Tynwald on Population Growth and Immigration*, 1979).

7. Cohen (1987) gives a superlative account of how difference is an essential component of identification as belonging to a closely bound Shetland community, which resonates strongly with what I came to understand of Manx village life.

8. For examples see Latour and Woolgar (1986: 285).

9. I would agree with Wynne (1992: 300) that Callon's theoretical approach does not take into account the historical, social, and cultural factors framing the relationship between fishermen and scientists from the fishermen's point of view in his insightful account of events in St Brieuc Bay (1986).

10. Strathern (1987) analyses the way concepts are used by anthropologists to describe the society of the 'other', creating awareness of different social worlds using terms belonging to their own.

11. This has parallels in recent trends in anthropological writing, for example see Clifford and Marcus (1986).

12. For a critical account see Emily Martin who points out that this approach tends not to recognise a world outside discourse, ignoring the structures of power that create the inequalities (1990: 72).

13. Concerning the identification of Symbolic Boundaries see McDonald (1987) on Brittany, Okely (1983) on gypsies, and Downey (1988) on nuclear scientists.

14. Issues concerning radioactivity did make regular appearances in the local press (the *Manx Independent*, the *Manx Examiner*): for example three pages were devoted to Chernobyl pollution in 'Bequereled Lamb' (*MI*, 8 October 1988, 7–9).

15. To see how information concerning the integrity of actors can have enormous consequences one need look no further than the role played by scandal in recent governmental changes, or the impact of the 'callgirl' scandal on the aspirations of the German nuclear industry.

16. On the island the processes involved are encapsulated in the local proverb of 'Manx Crabs': the crabs, thrown into a bucket, expend all their energy in pulling down any crab which looks as though it is going to manage to escape.

17. To quote Geertz: 'analytical dissolution of the unspoken premise from which common sense draws its authority – that it presents reality neat – is not intended to undermine that authority but to relocate it. If common sense is as much an interpretation of the immediacies of experience, a gloss on them, as are myth, painting, epistemology or whatever, then it is,

like them, historically constructed, and, like them, subjected to historically defined standards of judgement . . . It is, in short, a cultural system . . . it rests on the same basis that any other such system rests; the conviction by those whose possession it is of its value and validity' (1983: 75–6).

18. Ardener's writing on 'muted groups' stems from a concern with the way that women's 'worldview' can be rendered invisible by dominant male structures. It should be emphasised here that 'muted' population need neither be marginal nor isolated, very different 'publics' constituting most of society are muted in different ways in relation to science.

19. The fieldwork on which the analysis of Cumbrian farmers' responses to Chernobyl is based was carried out by Jean and Peter Williams, whose 'insider' status contributed greatly to the quality of material gathered.

20. The Island followed MAFF advice as to how long the contamination was likely to last (ranging from twenty-one days in the first few weeks, through two years to the present estimate of thirty-one years) and experienced the same disruption of lamb sales in the first year of monitoring.

21. See Boon (1982 Introduction) on how, at the nexus of cultural overlap, exaggeration results from process of identification, 'us' in opposition to 'them'; also Chapman (1982), McDonald (1987), Okely (1983) on how the experience of the meeting of two cultures results in misunderstanding that reify rather than dissolve stereotypes.

22. See Wynne (1992) on the way in which charges of 'fatalism' arise from the clash of cultural idioms, one of which assumes prediction and control to be the normal reponse to events, whilst the other emphasises flexibility in response to natural events.

23. It is interesting that nuclear installations are often located in 'remote areas' including the Celtic peripheries, where they are perceived to be 'out of place', leading the reflection on the nature of the locality as well as the technology concerned: see Chapman (1993), and Zonabend (1989).

24. The nine Tynwald departments replicate their Whitehall counterparts in as far as is possible. This chapter is concerned with the Department of Local Government and the Environment, and the Department of Agriculture and Fisheries.

25. The following remark concerning a local issue was typical of the approach taken towards elected representatives. A tirade of complaint wound up with the words: 'I'm not happy about this then and I'm going to have word about it with that Miles Walker the next time I see him.' None of the listeners doubted that this would be the case. Miles Walker is chief minister in the House of Keys.

26. Bourdieu describes the concept of habitus (1977: 78).

27. The island was actually having to deal with several difficult environmental issues as efforts were being made to bring pollution management into line with European directives.

REFERENCES

Ardener, E., 1989, 'The problem revisited', *The Voice of Prophecy and Other Papers* (Oxford: Basil Blackwell). (First published in *Perceiving Women*, edited by S. Ardener 1975, London: Dent)

Ardener, E., 1985, 'Social Anthropology and the Decline of Modernism', *Reason and Morality*, edited by J. Overing (London: Tavistock).

Barnes, B. and Edge, D. O., 1982, *Science in Context* (Milton Keynes: Open University Press).

Becher, T., 1989, *Academic Tribes and Territories* (Milton Keynes: Open University Press).

Beck, U., 1992. *The Risk Society* (London: Sage).

Boon, J., 1982, *Other Tribes, Other Scribes* (Cambridge University Press).

Bourdieu, P., 1977, *Outline of a Theory of Practice* (Cambridge University Press).

Bourdieu, P., 1981, 'Men and Machines', *Advances in Social Theory and Methodology: towards an integration of micro- and macro-Sociologies*, edited by K. D. Knorr-Cetina and A. Cicourel (London: Routledge and Kegan Paul).

Callon, M. 1986, 'Some elements of a sociology of translation: domestication of the scallops and the fishermen of St Brieuc Bay', *Power, Action and Belief: a new sociology of knowledge?*, edited by J. Law (London: Routledge, and Kegan Paul).

Chapman, M. 1978, *The Celtic Vision in Scottish Culture* (London: Croom Helm).

Chapman, M. 1982, '"Semantics" and the "Celt"', *Semantic Anthroplogy* (ASA Monographs 22), edited by D. Parkin (London: Academic Press).

Chapman, M. 1993, 'Copeland: Cumbria's Best Kept Secret', *Inside European Identities: ethnography in Western Europe*', edited by S. MacDonald (Oxford: Berg).

Clifford, J. and Marcus, G. E., 1986, *Writing Culture: the poetics and politics of ethnography* (Berkeley: University of California Press).

Cohen, A., 1987, *Whalsey: symbol, segment and boundary in a Shetland Island community* (Manchester University Press).

Douglas, M., and Wildavsky, A. 1982, *Risk and Culture* (Berkeley: University of California Press).

Downey, G., 1988, 'Reproducing cultural identity in negotiating nuclear power: the union of concerned scientists and emergency core cooling', *Social Studies of Science*, 18, 231–64.

Geertz, C., 1983, *Local Knowledge. Further Essays: interpretative anthropology* (New York: Basic Books).

Giddens, A., 1991, *The Consequences of Modernity* (Cambridge: Polity Press).

Herzfeld, M., 1987, *Anthropology Through the Looking Glass: critical ethnography in the margins of Europe* (Cambridge University Press).

Isle of Man Local Government Board, 1986, *Report to Tynwald: BNFL Ltd – Sellafield Operations*, Douglas.

Kermode, D. G., 1979, *Devolution at Work: a case study of the Isle of Man* (Farnborough: Saxon House).

Knorr-Cetina, K. 1988. 'The micro-social order: towards a reconception', *Action and Structure*, edited by N. G. Fielding (London: Sage).

Latour, B., 1986. 'The Powers of Association' in *Power, Action and Belief: a new sociology of knowledge*, edited by J. Law (London: Routledge and Kegan Paul).

Latour, B. and Woolgar, S., 1986, *Laboratory Life: the construction of scientific facts*, 2nd edn. (Princeton University Press).

McKechnie, R., 1993, 'Becoming Celtic in Corsica', *Inside European Identities: ethnography in Western Europe*', edited by S. MacDonald. (Oxford: Berg).

McDonald, M., 1987, 'The politics of fieldwork in Brittany', *Anthropology at Home* (ASA 25), edited by A. Jackson (London: Tavistock).

Martin, E., 1990, 'Science and women's bodies: forms of anthropological knowledge', *Body/Politics: women and the discourses of science*, edited by M. Jacobus, E. Fox-Keller, and S. Shuttleworth. (London: Routledge).

Okely, J., 1983, *The Traveller-Gypsies* (Cambridge University Press).

Paine, R., 1987, 'Accident, ideologies and routines: 'Chernobyl' over Norway', *Anthropology Today* 3, 4, August.

Paine, R., 1992, in Special Chernobyl Issue, *Public Understanding of Science* 1, 2.

Report of the Select Committee of Tynwald on Population Growth and Immigration, 1979 (Douglas: Tynwald Publications).

Strathern, M., 1987, 'Out of context. The persuasive fictions of anthropology', *Current Anthropology* 28, 3, 251–81.

Welsh, I., 1993, 'The Nimby Syndrome: its signficance in the history of the nuclear debate in Britain' *British Journal for the History of Science* 26, 15–32.

Wynne, B. and Smith, R., 1989, *Expert Evidence: interpreting science in the law* (London: Routledge).

Wynne. B., 1991, 'Knowledges in context', *Science, Technology and Human Values* 5, 16, 111–21.

Wynne, B., 1992, 'Misunderstood misunderstandings: social identities and public uptake of science', *Public Understandings of Science* 1, 281–304.

Zonabend, F., 1989, *La Presqu'île au Nucléaire* (Paris: Odile Jacob).

7 Authorising science: public understanding of science in museums

SHARON MACDONALD

Science communicators are widely acknowledged to have an important role to play in the public understanding of science. Despite this acknowledgement, the role is often seen as one of simply transporting information, like so many potatoes on a conveyor belt, from the world of science to the public (the transportation model) or else as a relatively straightforward matter of simplification and translation (the translation model). The aim of this chapter is to show that there is more to the communication of science than this supposedly value-free simplification and packaging; and that science communication involves selection and definition, not just of which 'facts' are presented to the public, but of what is to count as science and of what kind of entity or enterprise science is to be.[1] That is, science communicators act as authors of science for the public. They may also, however, by dint of their own institutional status, give implicit stamps of approval or disapproval to particular visions or versions of science. That is, they may act as authors with special authority on science – as authorisers of science.

Museums are one type of institution involved in the communication, authoring, and authorisation of science, and as such they share many of the same problems, and can exemplify many of the same issues, as other science media.[2] Although museums and science centres reach a smaller public than do some other science communicators – such as schools and television – this public is nevertheless substantial.[3] Over the last decade in particular, many museums, and their relatives the science centres, have been involved in developing new science communication strategies, strategies which have increasingly gone under the label of 'the public understanding of science'.[4] Although museums and science centres clearly have much overlap with other science media, we should note that there are likely to be significant differences of emphasis too. For example, the relative permanence of displays in science museums and science centres compared with particular lessons or television programmes, and perhaps their immediacy and the presence of 'the real thing', may give them a stronger authorial presence than those of other media.

Not only do science communicators define science for the public, they also in effect build a vision of 'the public', and the kind of 'understandings' that the public can be expected or hoped to make, into their communications. For example, the public may be assumed to be lacking in any scientific knowledge or as already well versed in its practical applications; it may be conceptualised as a large undifferentiated mass or as comprised of groups with different interests; it may be seen as bored or fascinated by science. The kinds of understandings which science communicators hope to further might be abstract and generalisable or context-specific; they might be about the social contexts of scientific work or about scientific facts; they might be for day-to-day use or knowledge for its own sake. While science communicators may set out to define precisely what they intend by each element in the 'public understanding of science', it is often taken as a relatively unproblematic label to indicate a shared enterprise. Even if definitions are made, there may well still be different interpretations involved in practice. These understandings of 'science', 'the public', and 'understanding' are difficult to see, however, without close observation of a science communication during its construction.

The focus of this chapter is the creation and reception of a major 'permanent' (i.e. at least ten years) exhibition in the Science Museum, London, an exhibition whose making I observed on a day-to-day basis for a year before its opening in October 1989.[5] The Science Museum is the main site of the National Museum of Science and Industry, and it acts in part as a record of high points of British achievement in science, industry, technology, and medicine. As such, it has a particularly significant status as an authority on science and related areas. My intention in this chapter is not only to look at the way in which science is represented in the final exhibition, however, but to investigate this in relation to both the making of the exhibition and its reception by museum visitors. By looking in detail at the construction of a science exhibition, my intention is to investigate both how and why science is represented as it is. As we shall see, the case-study – *Food for Thought: the Sainsbury Gallery* – is a particularly interesting one because its makers explicitly set out to challenge what they saw as more orthodox views and presentations of science. The story below, then, tells of the type of challenge which they made, how far they managed to carry it out, and what its visitors made of it.

Contexts and visions of 'public understanding of science'

Food for Thought was the first major exhibition in the Science Museum to be begun and completed under what was sometimes talked about in the Museum as a new era. Since Dr Neil Cossons became director of the Museum in 1986, there had been a good deal of change, much of it going under the Museum's new 'mission statement' of 'the public understanding of science'. The onus was on the group of Museum staff whose job it was to define the content of *Food for Thought* and to construct an exhibition which would exemplify a 'public understanding of science' approach. Before describing what they meant by this, let me briefly describe the context in which the phrase 'public understanding of science' had entered the Science Museum for, of course, authorship always takes place within a particular social, political, and economic context and, as we shall see, it is not in any case quite clear just who the author of *Food for Thought* is.

The phrase 'the public understanding of science' seems to have become prevalent in certain areas of public discourse in Britain in the wake of the Royal Society report on public understanding of science, published in 1985.[6] The Royal Society report was intended to argue the case for the importance of science – its relevance and use-value in modern society. As Bruce Lewenstein has argued of the use of the phrase 'public understanding of science' in the United States in the post Second World War period, what was frequently meant by its advocates was 'public *appreciation* of science'.[7] That 'science' is a body of knowledge produced by a fairly readily identifiable scientific community seems to be assumed in the report. The report is very specific, however, when it comes to 'the public', and much of its argument seems to be directed at one particular group of 'the public': 'people responsible for major decision-making in our society, particularly those in industry and government'.[8] As well as the decision-makers, the report classifies 'the public' as either 'workers' or 'citizens', and all are enlisted as members of 'the nation'.[9] In keeping with these definitions of the public, two main justifications are given for the national importance of science. These are, *economic* – 'national prosperity'; and *political* – the making of informed decisions within a democracy.[10]

Despite enlisting the public as citizens and decision-makers, however, the report works with an implicit model of the public as deficient and misguided in its present 'lack of uptake of science', a model which has been referred to as a 'deficit model'.[11] As the Introduction to this book suggested, this model is very much from the agenda of the scientific establishment. 'Science' is taken as a professionalised and distinctive domain that is, by definition, bounded off from lay people. The main way in which the perceived problem of public ignorance about science is to be dealt with, then, is by getting more science 'out' or 'across' that boundary to the public. It is not a model which questions whether, or in what ways, that boundary does or should exist in the first place. It also entails an implicit assumption that knowing more science will lead to greater public support or appreciation for science, though as research has indicated this certainly cannot be relied upon, greater knowledge about science sometimes leading to greater scepticism of it.[12]

In taking up what the director of the Science Museum has referred to as 'the brand name of public understanding of science',[13] the Science Museum was recognising a degree of overlap between its concerns and those of the Royal Society. The declining

status of British science and technology internationally was seen to have consequences for the status of Britain's National Museum. Was the Science Museum to
continue to try to represent both internationally significant and British achievements – achievements which were rarely as synonymous as they had sometimes
seemed to be in the past? If the Museum's role were to be based on the National,
then it faced confining itself largely to historical representation. However, particularly
given the Prime Minister, Mrs Thatcher's, widely reported negative views on
museums as being 'really rather dead',[14] such an emphasis was one which would be
unlikely to win the Science Museum much favour – or funding – in government
circles. The public understanding of science label enabled the Museum to make its
arguments in terms of a national *and* contemporary role. This did indeed seem to
appeal to the prime minister who provided a supportive letter to the Science
Museum's trustees, indicating approval of the stance which they were taking:

> I detect with appreciation [the Science Museum's] first steps to becoming not only the
> nation's showplace for the best in contemporary science and technology but its
> expanding role in promoting a broader public understanding of these important
> issues.[15]

As in the Royal Society report, the Prime Minister related the significance of
public understanding of science to the national economy: 'industrial success depends
on national attitudes to science, engineering and manufacturing. That is why I am
delighted that the Science Museum . . . displays the most modern technologies'.[16]

Like the Royal Society on behalf of scientists, then, the Science Museum was
using the public understanding of science brand-name in part to argue its case for
a share of the shrinking pool of government funding for public institutions. The
brand-name metaphor is an apt one, for it became indeed a kind of marketing label,
and, as the Science Museum's director observed, it could also be used to apply to
'products' which the Science Museum had long been offering.[17] However, the Science Museum was also having to sell its product to its visitors and potential visitors.
Visitors had come onto the museum agenda with a new urgency during the 1980s,
as the use of attendance figures as performance indicators on which funding might be
based loomed, and as more and more publicly funded museums introduced admission
charges. The Science Museum began charging for admission in October 1988. If the
Science Museum, like other public institutions, had long spoken in the name of 'the
public' and – like the Royal Society – had addressed that public as 'workers' and
'citizens' to be offered the useful and enlightening fruits of science and technology,
it was now faced with individuals with spending power who would make up their
own minds about whether to visit the Science Museum or go elsewhere. The
Museum, then, was perforce dealing with 'the public' not only as citizens, but also
as 'consumers', consumers who would make their own choices and who would not

necessarily relate to the Science Museum in the terms which that institution might have wished to set down. No longer could the Museum rest assured in the knowledge that what it was doing was good and important according to its own canon: visitors mattered very tangibly and would vote with their purses.[18] It was recognised throughout museums that much of the competition which they faced came from other uses of leisure time, such as staying at home watching television or visiting Madame Tussaud's. So, where the Royal Society had addressed the public as workers, the Museum found itself dealing with a public defined primarily as leisure-seekers. Of course, the Museum could try to resist these definitions, and during the period of my fieldwork there was much debate both within the Science Museum and throughout the museum world about this. To some extent, however, these definitions of the public were being authored outside museums themselves, and were being inscribed into funding arrangements.

None of this is to say that a definition of the public which included the idea of it as consuming and leisure-seeking was necessarily a bad thing. Many staff within museums welcomed the greater opportunity that this seemed to bring to orient exhibitions more towards visitors' wants and needs. As far as the public understanding of science equation was concerned, no longer was it about the view *of* the public *from* the vantage-point of science, but a more complex relationship in which the public's wishes and definitions needed to be taken into account. To put this in terms of our authorial metaphor, readers were not simply expected to be given what was good and edifying for them, but their own tastes and preferences needed to be accounted for, and written in to the communication product, too.

These definitions of the public as consuming and leisure-seeking are not, of course, the whole story. They were neither unanimously accepted by Museum staff, nor do they tell us all that is significant about the way in which the public is defined. We still need to know what kind of consumers and leisure-seekers the public is – spontaneous or careful? Purchasing entertainment or education? Homogeneous or segmented by race, class, and gender? And so forth. And we also still need to know what other definitions are brought into play in practice. Just as the definitions of 'the public' are configured and reconfigured by science communicators, so too are those other terms within their semantic field, namely 'science' and 'understanding'.

A new vision? Food for Thought

Before discussing the ways in which 'science', 'the public', and 'understanding' were used during the making of *Food for Thought* let me first give some background on the exhibition, its making, and the research on which this chapter is based. *Food for Thought* is, as its name implies, an exhibition about food. It is a large exhibition, covering $810m^2$, containing some 160 text and graphic panels and about 170 three-dimensional exhibits. It was also a substantial exhibition in terms of time and money. It took an exhibition team of six nearly eighteen months working full-time and often overtime to complete the exhibition, and in addition numerous other people were involved. It cost £1.2 million, excluding the Science Museum's own staff costs, £750 000 of which was provided by the main sponsor, or patron, the Sainsbury Family Charitable Trust, and a large proportion of the rest from a variety of other sponsors with an interest in the subject matter.

The research on which this chapter is based was an ethnographic study of the making of *Food for Thought* in the year leading up to its opening. As the ethnographer on the project, I spent much of my time with the six women museum staff who constituted the exhibition team. They were responsible for the majority of decisions about the content of the exhibition, for writing the script, for selecting the exhibits, and for generally managing all of the bits and pieces and people who play a part in making an exhibition actually happen. There were many more of these bits and pieces and people than I had ever imagined in advance; they included, to give but a few examples, historical objects, fake food, faxes (thousands of them), the budget, sponsors, picture researchers, designers, builders, and nutritionists. In addition to tagging along with the team, scribbling my notes in a corner, asking questions, and fetching the sustaining fizzy drinks and crisps as the hours became longer and more taxing, I also studied the stacks of paperwork that had accumulated in the 'Food' offices, attended exhibition-relevant meetings elsewhere in the museum, and interviewed staff in the Science Museum and at other museums and science centres. Once the exhibition was opened, a sample of visitors to it was observed in the gallery and interviewed at length, and this data is the basis for the discussion of visitors to *Food for Thought* below.

The team clearly had a strong authorial role in the creation of the exhibition, though, as I have noted, it had numerous other players to deal with, others who played a vital role in enabling the exhibition to happen, but who also made resistances of various sorts. Each exhibit in this large exhibition has a story of its route into the displays behind it, and these stories often reveal quite complex plots, so the account below is only intended to give some examples of the kinds of interventions that sometimes interfered with the team's attempt to challenge more orthodox science presentations. In other words, below I attempt to show just some of the more note-worthy ghost-writers who were involved in the creation of *Food for Thought*.

During the making of *Food for Thought* there was much talk, particularly in the team offices, of challenging established modes of museum science representation. A rhetoric of newness and difference from what has gone before may well, of course, be typical among creators of an exhibition, because it is through such notions of discontinuity and difference that creativity is principally established in dominant Western cultures.[19] Although there is doubtless an element of exaggeration and stereotyping involved in the contrasting of traditional representations and the new exhibition, it is through such differentiation that the team defined their own objectives and ideals. It may also be the case that this contrast was particularly strong because of the sense of change wrought by the presence of a new director, and because the team members themselves had a strong sense of being non-traditional representatives of the Science Museum in many ways. The team referred to its approach to the exhibition as 'public understanding of science', interpreting this as involving primarily a foregrounding of the public, the museum visitors.

Although the vision of visitors as consumers and leisure-seekers was central to the team members' definition, their approach also involved questioning the relationship between the public and science. Just as the Royal Society and the Science Museum had adopted the idea of 'public understanding of science' within a particular set of interests, so too did the team. In part these concurred with those of the Science Museum – they were employees within it, and charged with one of its particularly prestigious tasks, after all. However, members of the team also assimilated the notion to their own professional and personal positions within the Museum, identifying *themselves* with a lay-public not to be cowed by science. This was something which was partially fostered by a new exhibition-management strategy which took exhibition-making out of collections-based curatorial departments into a new division of 'Public Services' set up specifically to manage the directly public-focused aspects of the Museum. While team members certainly had considerable curatorial expertise in their own areas, they were not selected to work on *Food for Thought* for these skills so much as for their likely abilities in communicating with the public. Not to be a subject expert was seen by them, on the whole, as a strength rather than a weakness in creating this kind of public understanding of science exhibition. Stories were told among the team of exhibitions which were basically curators' Ph.D. theses pasted up onto panels, of exhibitions in which the curators were 'out of touch'[20] with the public, of exhibitions which were written for 'the three other experts in the field'. *Food for Thought* was not to be like this.

During the making of the exhibition it was common for team members to 'stand in for' the public, by trying out on each other the text they had written, for example. Dialogue like the following was not at all unusual:

ANN Does this make sense? Do you *really* think it's clear enough?
HEATHER Well, I understood it – and if I can understand it anybody can.

This self-identification with, and even celebration of, a fairly ignorant lay-public for whom science was often difficult and distant was also, I suggest, fuelled by the team's own status and gender within the Museum. Team members were relatively low-graded to be charged with the task of making an exhibition (most were on temporary promotions), something which they related in part to not having entered

the Museum as subject experts. To create a team of six women was unprecedented
and not purely fortuitous in an institution where women curatorial staff constituted
only 30% of the total (and those at disproportionately lower grades). Their gender
and status as 'non-experts' seemed to become meshed together as markers of difference from what they mostly conceptualised as a traditional and conservative museum
establishment in which 'science' itself was masculinely gendered. They sometimes
talked of the paucity of representations of women in the Science Museum; and they
described examples of sexism which they had encountered, both personally and in
relation to the exhibition. What would *Food* be 'but a few ovens and a load of old
cookery books?', one male Museum official was reputed to have said. Their identification with the public, then, was an identification with a public which had been
disregarded by the museum and scientific establishment; and part of the rhetoric
during the making of the exhibition was not just about getting more science to that
public, but about challenging some of the establishment high ground. The team
enjoyed the consternation which it caused in some quarters of the Museum by the
rumours which it started about a (fictitious) large lump of cheese complete with
mouse which would hover over the exhibition, or the (again fictitious) giant tea-cup
which would dominate the Museum's central atrium, or the (not fictitious) recon-
structed McDonalds and enormous chocolate mousse pot that would be included in
the exhibition.

In contrast to those exhibitions which were written for curators rather than visitors, exhibitions which were said to 'go over people's heads', *Food for Thought* was
to be 'accessible' and 'user-friendly'. The users or public were defined by the team
in a number of ways. Firstly, they were to be lay-people with little or no prior
knowledge of school-type science and probably with a fear of science. Here again,
there was often self-identification by team members, the majority of whom had
degrees in history or archaeology rather than science. However, the idea of the public
as having little formal science, and regarding science as difficult and distant, was
not the deficit model described above, for the team also saw the public as having
relevant practical knowledge that, although not always classified as science, could be
drawn upon in helping visitors' understandings. Science, then, was not to be just
formal science or associated with a specialist community or sophisticated technology,
it was also to be related to everyday practical experience.

Although the exhibition was specifically designed for 'family groups', in fact this
often seemed to mean children.[21] The 'reading-age' of the main text, for example,
was planned as being for not higher than 11-year-olds. There was great emphasis
on creating an exhibition which would be 'fun'; and one of the most dominant
features of *Food's* imagined audience seemed to be that it would be in need of entertainment, boredom being seen as the state into which visitors would readily slip
unless constantly kept occupied by the exhibition. A recurring theme in the team

members' discussions was whether the gallery would be 'boring' or 'busy' and 'interesting' and 'fun'. Names for exhibits, particularly during the early stages of making, were often from the world of childhood and cartoons; and the great majority of names for the exhibition produced during the team's own 'brainstorming' were a good deal more jokey, childish, and fun than the exhibition's final title. An important aspect of entertaining visitors was the use of hands-on interactive exhibits, of which there are a fairly large number in *Food for Thought*. Interactives were also thought to be educational, of course, and again they present a particular vision of science.[22] In interactive exhibits visitors are encouraged to become involved, rather than be passive viewers, and this was explicitly seen by the team as a further element of removing some of the barriers around science. They also planned to avoid as far as possible having barriers around any of their exhibits, again projecting their ideas about the relationship between the public and science deeply into the final presentation.

In their quest for entertainment, interest, and fun, visitors were conceptualised as active choice-makers. This was both one of the storylines presented – the expansion of food choice in Britain over the century with improved transportation and so forth – and an idea built into the structure of the exhibition. In terms of the exhibition structure, there are various routes which can be taken, choices of where to go and what to see made inevitable by the plethora of exhibits and the many different areas of the exhibition. This idea of choice was also extended, to a degree, into science: diet was to be presented as a matter of individual choice rather than as a set of rules, and different sides of arguments (over, say, fat in the diet), and different sets of guidelines (for example, on calorie intake) were to be presented. The team members tried to avoid being prescriptive, though they did sometimes interpret 'public understanding of science' as entailing giving guidelines that would be useful to the public in their everyday lives.[23] But, in the controversial area of nutrition in particular, visitors were to be invited to make up their own minds – to make choices even in domains of expert knowledge.

In team discussions, the public was often talked about as 'everyone'. Team members generally seemed to assume that there would be few exceptions to their vision of a scientifically uninterested and even fearful lay-public. At another level, however, like the Royal Society and the Science Museum, the team also equated this public with the nation. However, this was not a nation of homogeneous citizens or consumers, but a nation segmented by ethnicity, class, and gender. This is illustrated, for example, in an exhibit consisting of a set of larders, each dated to coincide with a major immigration influence on Britain. Again, it was felt by the team that just to include some of the groups which did not usually find any place in the Science Museum – even if they were being represented as 'users' rather than producers of science – was to challenge some of the orthodoxy. For the same reason, photographs and models often included women.

The public, then, was conceptualised as largely fun-loving and technophobic, as choice-making, but easily bored, as child-like, British, and (sometimes) of a particular race, gender, and class. So what kinds of 'understanding' were expected of it? There seemed to be little expectation that understandings would be at all sophisticated, and quite often during the making of the exhibition ideas were abandoned because they were deemed too difficult to present, particularly within a restricted space. Exhibition-strategies designed to make science accessible, such as that of beginning each topic area with something with which visitors would be familiar (for example, a supermarket checkout desk), and the aim to make text short and simple, precluded the opportunity to delve far into the unfamiliar. *Food for Thought* does not for the most part attempt to convey abstract principles, but focuses instead upon information within its context of application or use. The kinds of understandings that *Food for Thought* seems to seek to develop are not so much a set of themes closely woven together, each expanding upon one another, as loosely related pieces of information among which visitors can choose and build together their own stories. It is significant that in the 'messages' around which the exhibition was constructed, the direct form of address (the pronoun 'you') is used almost invariably: in the arena of understanding, the visitor is to piece together a story incorporating their own prior experience and practical knowledge. Perhaps more than anything, the understanding that is being fostered here is that the boundaries between the familiar and the unfamiliar, between the everyday and science, can be crossed.

Just as the team conceptualised the public, it also defined science. Although the team members generally seemed themselves to see it as something rather dull, they were determined to turn it into fun and to break down some of the barriers surrounding it. Science was not to be abstract and specialised, but was to stretch in a continuum into the everyday and practical. The team also, at times, sought to question the nature of scientific knowledge by, for example, showing variation in different national calorie-intake recommendations, or by noting areas of disagreement among scientists. However, these were not the only visions of science. In their aim of allowing the public to make choices, the team members would talk about their own role as simply presenting the facts in a value-free way: 'we don't make value judgements' was a phrase which came up several times, though this was on at least one occasion acknowledged to be 'impossible of course'. Therefore, although the exhibition contains instances where the status of scientific knowledge is questioned, for the most part 'facts' are presented unproblematically. This should not be surprising, of course, for this is an exhibition about food rather than the status of scientific knowledge itself, and is aimed at a lay-audience. To dispense with facts altogether,

or to present all of the scientific or indeed historical knowledge on display as contentious, would neither have been an accurate reflection of the epistemological status
of much of that knowledge, nor would it have provided much scope for talking about
the subject-matter in hand (problems all too familiar to post-modern ethnographers).
There were further resistances that were made in the name of science however, and
it is to these resistances – resistances which played a role in the shaping of the
exhibition – which I now turn.

Authors, resistances, and ghost-writers

Within the Science Museum, *Food for Thought*, like other exhibitions, was identified
with the team or individuals responsible for planning its content and writing the
script. Indeed, Museum staff often refer to exhibitions by the name of their makers
rather than the title or subject-matter of the exhibition. Despite this internal recognition of authorship, however, the team is not the official author of an exhibition.
Indeed, team members' names appear nowhere in the final exhibition, though sponsors, designers, the academic advisory panel, and those who gave assistance or
objects, all get a mention. The reason given for this 'deletion' is that the exhibition
is not the product of the team, but of the Science Museum. The team is subsumed
within the institution. The official author of the exhibition, then, is the Science
Museum itself.

This is not, however, simply a matter of relabelling the end-product, of the Science Museum as an inconsequential *nom de plume*. The fact that the exhibition is
to be a Science Museum exhibition – that it is to speak in its name, under the mantle
of its authority – has an effect upon decisions made during the making of the exhibition. There are also managerial implications, for the exhibition team must 'report
up' about its plans for the exhibition, and its superiors may well make suggestions
and could, feasibly, stop an exhibition going ahead if it were deemed to be inappropriate. In the case of *Food for Thought*, the authorial presence of the Science Museum
was felt, not so much through the hand of management, as through the awareness,
both on the part of the team members themselves, and others with whom they dealt,
that this was an exhibition which would have to be worthy of a place within the
Science Museum's authorising dominions. A particular vision of science, and of the
Science Museum, as a particular kind of authority on science, found its way into
the making of the exhibition. The following examples illustrate some of the main
areas from which the challenge to traditional visions of science that the team had
talked about was tempered.

As the exhibition progressed, the team seemed to have some recourse to more
orthodox, less social, representations of science, representations justified or seen as
inevitable because of the fact that *Food for Thought* was to be a Science Museum

exhibition. Here, a contrast was sometimes made with museums of social history, and a distinct identity and role of *science* museums accepted. Part of the reason for this was that, just as the team's vision of the public needed to bear relation to the actual visiting public, so too did its vision of science, and of its own institution's identity and role, need to be recognisable at least in part to the exhibition's various contributors and audiences. These included sponsors, advisors, and people asked for assistance of one sort or another. For the team members to speak as the Science Museum was an important statement of their own legitimacy to act as communicators of science; it provided them with authority in their dealings with the various others. Were their exhibition to be unrecognisable as a Science Museum exhibition – were it to be thoroughly questioning the whole idea of science as a privileged way of knowing, for example – the team's power to enlist the assistance of others would have been weakened. This is not to say that the team necessarily wanted to question the authority of science to this extent. Although I have described a context in which the team wanted to reconfigure the authority relations between science and the public, this did not necessarily entail divesting science of its authority altogether. The team was after all, composed of employees of the Science Museum.

The authority of the Science Museum and of science also tended to emerge when there were competing interests or particularly awkward dilemmas during exhibition-making. At such problematic conjunctures, the fact that the exhibition was 'a Science Museum exhibition' provided a readily available and relatively uncontestable rationale for the decisions made. One particularly dramatic example of this was a major editing of the exhibition – carried out in response to questions from the Museum's Director about the 'clarity of the messages' – which resulted in the central 'Food in the Factory' sections of the exhibition becoming largely devoid of the social, political, and historical dimensions which had previously been intended. The focus almost exclusively upon technological aspects of production was justified, albeit with misgivings from some team members, with the statement: 'we are the Science Museum after all'.[24]

It was not only the team members who shaped the exhibition through their relations to the idea of the Science Museum. So too did other groups with which they dealt. A clear example here was the conservative effect of perceptions of the Science Museum upon the team's intentions to depict disagreement among scientists. The team members had planned, in their section on 'Food and the Body', to include a set of different statements by contemporary nutritionists to illustrate the disagreement among them. The exhibition includes a panel entitled 'Consensus or controversy?' Nutritionists known to hold quite divergent views were asked to provide a short statement of their views, together with a mugshot. The statements which they provided, however, far from illustrating disagreement, were all rather general and far from controversial. Perhaps in the knowledge that their opinions would be on display for at least ten years, and that these were views to be included in that monumental record, the Science Museum, the scientists played safe.

These examples show, then, that the very fact of being the Science Museum, invested as it is with the rather special authority of science, has an effect upon the exhibition. The examples given are all quite noticeable ones, but the effects are felt in many smaller ways too. None of this is to say that all Science Museum exhibitions will turn out identical, or that the effects will necessarily be felt in quite the same way, for a dialectic is set up between all of the various visions and players involved, players which include the physical objects, the budget, visitors, and so forth. We cannot ignore, however, the feedback system whereby perceptions of what the Science Museum has displayed and what it is seen to stand for are played back into new exhibitions. To make claims about what would or would not be appropriate for an exhibition of this sort, and in particular to appeal to 'scientific accuracy', was a significant authorising device. It was not only used by the team, but also by other interested groups. For example, the advisory panel of nutritionists led the team members to alter their early plans for the whole framework of the exhibition through their claims that the nutritional principle that the organisation had been based upon was now considered misguided. Later on, when the team members sent their scripts to the panel and to their sponsors for comments on their 'factual accuracy' the latter, together with the idea of what was 'scientific', sometimes became a contentious arena around which to negotiate other interests.

Reading 'science'?

I have suggested that the visions embodied in the rhetoric during the making of *Food for Thought*, especially during its earlier stages, were tempered to some extent as the exhibition progressed by pre-existing definitions of science and of the perceived roles of the Science Museum. This was by no means a complete shift, but it left the exhibition with variable depictions of science and of public understanding of science; and with what, to me at least, seemed like an exhibition less different from previous ones than talk during its making had led me to imagine. Nevertheless, the intentions to create a visitor-centred exhibition in which 'science' was predominantly presented as located in the everyday and familiar, and as interactive and easy, still came through into the final exhibition. But how did its visitors see it? Here I focus on the question of visitors' conceptions of 'science', and in particular whether they felt that *Food for Thought* challenged their preconceptions. I look too at the question of authorship – of who the exhibition's visitors thought had created the exhibition and any consequences they saw to follow from this.

The study of audiences involved tracking groups of visitors through the exhibition and then interviewing them at some length, in the groups in which they visited, after their visit.[25] These interviews consisted of open-ended questions designed as far as possible to encourage visitors to talk about the exhibition in their own terms; and they were recorded. In the first part of the interview visitors were asked simply to describe their visit to the exhibition and to tell the interviewer what they thought its theme was. Their replies rarely mention science or technology. Instead, as I have described in detail elsewhere, they are more likely to reconstruct the exhibition into narratives about 'the history' or about 'healthy eating'.[26] Both of these themes are present in the exhibition, but what is significant about the visitors' accounts is that they link together exhibits which were not linked either on the ground in the exhibition or in its conceptual framework. Significant too is that 'healthy eating' is frequently described as 'good foods and bad foods' or 'what you should and shouldn't eat' even though the exhibition-makers had tried, in part at least, to avoid being prescriptive or making this kind of classification of foods. Indeed, the exhibition states directly, though not particularly prominently: 'Most scientists agree that no one food in isolation is "good" or "bad".'

Although visitors seemed rarely to mention science and technology spontaneously, we asked them directly whether they regarded *Food for Thought* as 'scientific' ('Does it strike you as a scientific exhibition?'). Replies revealed not just visitors' thoughts about *Food for Thought*, but their ideas about what might be meant by 'scientific'. To simplify somewhat, replies could almost all be classified as follows. A small minority of visitors held the view that science was 'about physics and things like that', and that by no stretch of the imagination could *Food* be regarded as 'scientific'. Another minority seemed to find that it was unproblematically 'science' because they readily accepted the view that science is 'in the everyday things too'. The replies of other visitors, however, revealed ambivalence and negotiation. They often replied to the effect that it had not struck them as a 'scientific' exhibition – and here they identified science as difficult and distant as anticipated by the exhibition-makers – but that they recognised that it was, or must be, in part or at some level. Discussion among visitors showed them working through the ways in which *Food for Thought* could be scientific, a process which often involved making distinctions between different 'levels of scientificness', or between 'real science' or 'pure science' and 'popular science'. What seemed to be going on here was that visitors found their preconceptions challenged by the exhibition and in response they began to rethink those definitions. During this process, science was 'partitioned' and, in the invocation of ideas such as 'pure' or 'real' science, 'science' proper was sometimes still moved elsewhere, still out of the public's grasp. Nevertheless, the exhibition seemed to have some power to challenge preconceptions of 'science', and this authority was one which was drawn in large part from the location of the exhibition: several visitors replied that the exhibition must, of course, be scientific because 'it's in the Science Museum'.

But who did visitors see as the author of the exhibition? While the Science Museum was a significant aspect of the public's framing of the exhibition, neither the Museum nor its staff were usually identified as its 'writers'. The most predominant answer to our question 'Who do you think wrote this exhibition?' was 'Sainsburys' or 'Mr Sainsbury'. The cue that visitors picked up on here was the title of the gallery – *Food for Thought: the Sainsbury Gallery* – and the Sainsburys reconstruction at one of the exhibition's entrances. That visitors regarded a food industry company as the writer of the exhibition might, presumably, raise questions as to the reliability of the information displayed: might it be biased?[27] Although some visitors did suggest that there were aspects of the exhibition which might be biased because, as they thought, it had been written by food industry representatives (for example, the omission of any mention of meat-stripping was pointed out by one) many seemed to assume that 'bias' would take the form of fairly obvious 'advertising', and that it would simply be a matter of the presence of the sponsor's name in the exhibition. In other words, they seemed to regard it rather like sponsors' names on footballers' shirts, rather than as something which might have considerable consequences for the content of the exhibition itself. Given that they saw sponsorship in this way, it is not surprising that most assumed that the 'advertising' involved was a fairly superficial matter which they would detect with ease and decide whether to buy or not. The following is just one example of this kind of reasoning:

Question: Do you think the fact that it's sponsored makes any difference?

  Answer: Well, I suppose it's an advertisement for Sainsburys ... probably because people think ... Well, food and Sainsburys are synonymous. But I shouldn't think it has a bad effect, no. I would say it's a fairly neutral effect. I don't think people are going to suddenly rush out and buy all the Sainsbury's things because they've seen it. So that way it's not necessarily an advertisement.

This general (though not unanimous) notion, that authorship by the food industry would only entail the industry giving its name and logo what prominence it could, might seem to indicate that visitors regarded the content of the exhibition as fair and uncontentious. Indeed, many made comments to this effect. Part of this, however, seemed again to relate to the authorizing power of science and of the Science Museum (and also, the perceived respectability of Sainsburys). When we asked about bias, visitors sometimes replied that 'scientists' would surely be involved, or that the Science Museum would presumably have 'its watchdogs'.

Various other chapters in this book examine the issue of the public's understandings of science and of scientific experts, and they illustrate both the *activity* of the public's knowledge production – something very much in evidence here in visitors' reconstructions of the exhibition – and the often careful and sceptical judgements which the public may make within specific contexts. The example of sponsorship here might suggest that the majority of the public does not ask particularly searching questions about the construction of scientific knowledge: they do not ask questions about silences, specific juxtapositions, and possible 'underlying' messages. However, we must remember context here too: we are dealing with family groups on a day out. Critical dissection is not the operative context. What is more, however, a key part of the context is the perceivedly benign and neutral nature of a public institution such as a national museum: it is not a context in which the authority or balance of the content is popularly perceived to be at issue.

Conclusions

This chapter has looked at issues surrounding the authorship and authorising of science communications. However, the aim has not been to celebrate the authorial role of those who might think of themselves as the authors, but to show how the authorial presence of others, including 'science' itself, affects the final 'text' and the readings of it. All definitions of 'science', 'understanding', and 'the public' take place within a specific social and political context of which its users may be more or less aware. Certain definitions, however, have more power to make themselves felt than do others. The case-study points particularly to the definitional significance of the institutional locus of the public representation. This is a pre-existing reality for the science communicators, and feeds itself back into the reality which they are trying to create in all kinds of ways. What is more, the very fact that a communication is for 'the public', and that it embodies a specific vision of that public, shapes the kind of representation made.

Studies of finished museum displays have frequently illustrated the hand of 'the state' at work in the representations. Control of the public, and interests in maintaining a particular status quo of class, gender, or ethnicity have all been described.[28] While the case-study in this chapter clearly shows routes by which 'state' interests make their way into communications intended for public consumption, it also shows that these routes are neither straightforward nor simple, and that groups and individuals with their own, possibly alternatively politicised, visions may deflect and redefine some of those interests. Together with recognition of the roles that the many other human and non-human actors may play,[29] we are left, then, with a more heterogeneous and complex picture of the processes involved in creating a science communication or museum exhibition. However, despite the contingencies, this is never a fully random or unpatterned complexity, for some actors are imbued with greater authority than others. The story of this chapter has been one of a relatively radical-popular vision of 'public understanding of science' finding itself rewritten – by no means wholly, but to some extent – by powerful ghost-writers; and finding itself read – again, by no means wholly – by visitors whose frames of reference are often drawn from alternative contexts.

A number of more specific comments can also be made about 'public understanding of science' strategies. A key aim in this particular exhibition, as in other 'public understanding of science' communications, was to make science *accessible* and to do so through using simple language and through presenting science through familiar contexts. While the visitor study clearly showed many visitors favourably contrasting this with other exhibitions said to be 'full of incomprehensible jargon', it does not necessarily follow that they came away either with a better understanding of science, or more empowered to deal with it. Even a visitor who takes on board the idea that 'science is in everyday things too', as one put it, may still shift 'real science' – difficult, formal, asocial science – elsewhere. Two other dangers may follow from the representation of science as familiar and everyday. The first is that, although the aim is generally to move beyond this and into more unfamiliar terrain, in practice, amidst constraints of space, word-limits and so forth, the representation may well not do so, or do so very little. In other words, it may end up saying very little that is new to its audience, something suggested by some of *Food for Thought*'s visitors. The other and greater danger is that the strategy may actually lead visitors into *not* asking questions, into a sense of security that the world of science *is* familiar; that factory production *is* essentially like that which goes on every day in the home; that there are no barriers around the big business worlds of food technology and distribution. In other words, the domestication involved in strategies of familiarisation and accessibility may act as a kind of intellectual narcotic. Might visitors have had more challenging questions to ask about the role of the food industry and about the content of the exhibition if it had included a realistic representation of, say, meat-stripping? Science can be difficult and distant, it can be gendered and racist, it can be hedged about with all kinds of barriers and vested interests. Understanding this is 'public understanding of science' too; though the case-study suggests that enlisting support to promote such understandings as these would not be easy within the contexts within which a national museum must operate.

Similar questions might also be asked of the theme of 'choice', a theme related to both the content and the structure of the exhibition. Like accessibility, 'choice' was seen as essentially democratic, allowing the visitors rather than the scientists or exhibition-makers to make the final decisions. However, choices had, of course, been prefigured, so begging the question of what the choices were to be made between. This message of choice, however, like that of familiarity, could give visitors a sense of security that the representation was inevitably balanced and fair, and again lead to a less, rather than more, critical approach.

These reflections are not intended to suggest that science communicators should abandon attempts to be accessible or to offer more than single accounts or answers.

They suggest, however, that neither accessibility nor choice is necessarily on its own a solution to 'public understanding of science', and that more questions about context, about authority, and about types of understanding still need to be asked. I should emphasise that these questions have come out of the process of going through visitors' words for many hours, reading the reviews, looking at and thinking about the final exhibition in relation to its formative ideas and enthusiasms, and listening to criticisms made of the exhibition by the team members themselves (criticisms which I am sure were harsher and deeper than those which any exhibition-makers satisfied with the maker's right to dictate display would have been), and after thinking about questions raised by others and by debates in other academic contexts.[30] They are questions which might seem unfair given the extent to which *Food for Thought* did challenge many expectations. However, it is important, I think, that we do raise such questions, partly so that the communication strategies that have become labelled 'public understanding of science' do not simply become part of a new unselfconscious canon. Just as we need to analyse what we mean by 'public', 'understanding', and 'science', we need also to see what particular assumptions, visions, and strategies – not all of which necessarily have compatible outcomes – have come to nestle under the 'public understanding of science' brand-name itself.

ACKNOWLEDGEMENTS

This research was directed by Professor Roger Silverstone, Media Studies, Sussex University, to whom special thanks are due both for his role in the research and for insightful comments on this particular piece. The research was supported by the Science Policy Support Group and Economic and Social Science Research Council under their 'Public Understanding of Science' programme; and was carried out at the Centre for Research into Innovation, Culture and Technology (CRICT), Brunel University. Considerable thanks are also due to the staff of the Science Museum, London, especially the now 'ex' Food Team, for their immense forbearance and hospitality, and for comments on this chapter. I would also like to thank Alan Irwin and Brian Wynne for a generous mix of constructive criticism, patience, and encouragement; and Gordon Fyfe, Keele University, for helpful discussion. Problems in the chapter remain, of course, my own.

NOTES

1. Cf. Silverstone, R., 'Communicating science to the public', *Science, Technology and Human Values* 16, 1 (1991) 69–89. This article includes brief descriptions of early results from three 'Public Understanding of Science' projects which concerned science communication.
2. The extent to which they may be similar or different from other media is discussed in Silverstone 'Communicating science', and also in Silverstone, R., 'Museums and the media: a theoretical and methodological exploration', *International Journal of Museum Management and Curatorship*, 7 (1988) 231–42; Silverstone, R., 'Heritage as media: some implications for research', *Heritage Interpretation Volume 2: the visitor Experience*, edited by D. Uzzell (London: Frances Pinter, 1989); Silverstone, R., 'The medium is the museum: on objects

and logics in times and spaces', *Museums and the Public Understanding of Science*, edited by J. Durant (London: Science Museum, 1992); and Morton, A., 'Tomorrow's yesterdays: science museums and the future', *The Museum Time-Machine*, edited by R. Lumley (London: Routledge/Comedia, 1988) pp. 128–143.

3. Figures on the visiting of science museums specifically are not compiled. However, research on museum visiting in general suggests that some 68% of the population has visited a museum in the last four years Merriman, N., 'Museum visiting as a cultural phenomenon', *The New Museology*, edited by P. Vergo (London: Reaktion Books, 1989).

4. For description and discussion of some of these see J. Durant (ed.) *Museums and the Public Understanding of Science* (London: Science Museum, 1992).

5. A fuller account of the making of the exhibition can be found in Macdonald S., and Silverstone, R., *Food for Thought – the Sainsbury Gallery: some issues involved in the making of a science museum exhibition* (Report, Centre for Research into Innovation, Culture and Technology, Brunel University, London, 1990).

6. The Royal Society, *The Public Understanding of Science*, (London: Royal Society, 1985).

7. Lewenstein, B., 'The meaning of "public understanding of science" in the United States after World War II', *Public Understanding of Science* 1,1 (1992) 45–68.

8. Royal Society, *Public Understanding*, p. 9.

9. I should note that this is my classification of a larger number of categories actually employed in the Royal Society report. The report's sub-division is as follows: '(i) private individuals . . . (ii) individual citizens . . . (iii) people employed in skilled and semi-skilled occupations . . . (iv) people employed in the middle ranks of management and in professional and trade unions associations; and (v) people responsible for major decision-making in our society, particularly those in industry and government. (Royal Society, *Public Understanding*, p. 7).

10. Ibid. *passim*.

11. See Wynne, B., 'Knowledges in context', *Science, Technology and Human Values* 16 (1991) 111–121; and Ziman, J., 'Public understanding of science', *Science, Technology and Human Values* 16 (1991) 99–105.

12. Wynne, B., 'Public understanding of science', in *Handbook of Science and Technology Studies*, edited by S. Jasanoff, G. E. Markle, J. C. Peterson, and T. Pinch (London and Beverly Hills: 1995) pp. 457–79.

13. Comment made during opening address to a conference on 'Museums and the Public Understanding of Science', (London; Science Museum, April 1992).

14. Comment made by Mrs Thatcher at the opening of the Design Museum, London, July 1989; reported in, for example, *Observer* 23 July 1989.

15. Letter reprinted in *Science Museum Review* (London: Science Museum, 1987), p. 5.

16. Ibid.

17. Neil Cossons, spoken comment, April 1992. For histories of the Science Museum, and its relationship to its public, see Day, L., 'A short history of the Science Museum', in *Science Museum Review* (London: Science Museum, 1987) 14–18; Bedini, S., 'The evolution of science museums,' *Technology and Culture* 6 (1965) 1–29; Butler, S., *Science and Technology Museums* (Leicester University Press, 1992).

18. See also Macdonald, S., and Silverstone, R., 'Taxonomies, stories and readers: rewriting the museums' fictions', *Cultural Studies* 4,2 (1990) 176–91; and Macdonald, S., 'Un nouveau "corps des visiteurs": musées et changements culturels', *Publics et Musées 3* (1993) 13–27.

19. See, for example, Gewertz D., and Errington, F., *Twisted Histories, Altered Contexts* (Cambridge University Press, 1992).

20. Unless otherwise specified inverted commas signify terms or phrases used by the team.

21. On the theme of the visitor as child see Silverstone, R., 'Caff society on a voyage of discovery', *Times Higher Education Supplement* (12 June 1992); and Macdonald, S., 'Un nouveau'.

22. For some discussion of this see, for example, Saunier, D., 'Museology and scientific culture', *Impact of Science on Society* 152 (1989) 337–53; and Macdonald, S., 'Cultural imagining among museum visitors', *Museum Management and Curatorship* 11 (1992) 401–9.

23. Macdonald, S. and Silverstone, R., 'Science on display: the representation of scientific controversy in museum exhibitions', *Public Understanding of Science* 1, 1 (1992): 69–87.

24. Macdonald S. and Silverstone R., 'Science on display'.

25. The visitor study was devised by Roger Silverstone, Gilly Heron, and myself and largely carried out by Gilly Heron, who also carried out preliminary analysis of the data. A fuller report of the visitor study is Macdonald, *Museum Visiting: a Science Exhibition Case Study* (Department of Sociology and Social Anthropology, Keele University, Working Paper, 1993).

26. Inverted commas in this section indicate terms or phrases used by visitors unless otherwise specified. For more on these themes see Macdonald, S., *Museum Visiting* and Macdonald S., 'Consuming science: public knowledge and the dispersed politics of reception among visitors', *Media, Culture and Society* 17 (1995) 13–29.

27. For a discussion of sponsorship and bias in museum exhibitions see Kirby, S., 'Policy and politics: charges, sponsorship, and bias', in *The Museum Time-Machine*, edited by R. Lumley (London: Routledge/Comedia, 1988) pp. 89–101.

28. For some examples of analyses which make such illustrations particularly well, though often subtly, see Bennett, T., 'The exhibitionary complex', *New Formations* 4 (1988) 73–102; Haraway, D., 'Teddy bear patriarchy: taxidermy in the Garden of Eden, New York City, 1908–1936', *Primate Visions*, Haraway D. (London: Routledge and Kegan Paul, 1989); Hewison, R., *The Heritage Industry* (London: Methuen, 1987); and various chapters in Lumley R. (ed.), *The Museum Time-Machine* (London: Routledge/Comedia, 1988).

29. The roles of physical non-human actors have not particularly been emphasised in this paper. For an attempt to classify some of the types of roles which such actors may play see Star, S. L., and Griesemer, J. R., 'Institutional ecology, "translations" and boundary objects: amateurs and professionals in Berkeley's Museum of Vertebrate Zoology, 1907–1939', *Social Studies of Science* 19 (1989) 387–420.

30. In particular, they have been informed by my own similar problems in writing ethnography, problems which I have discussed in Macdonald, S., 'Anthropology dangerously close to home: some problems of ethnography in a parallel context' (unpublished, M.S., 1991). See also Macdonald, S., 'The museum as mirror' in *Anthropology and Representation*, edited by A. Dawson, J. Hockey and A. James (Routledge, forthcoming). There is a considerable literature in anthropology now on these 'post-modern' issues: for example Clifford, J., and Marcus, G., *Writing Culture* (Berkeley, University of California Press, 1986); Marcus, G., and Fischer, M., *Anthropology as Cultural Critique* (Chicago University Press, 1986).

8 Nature's advocates: putting science to work in environmental organisations[1]

STEVEN YEARLEY

Environmentalism and the public's understanding of science

Since the late 1980s Britain has experienced a rapid rise in popular and media interest in environmental matters. The level of public concern with green issues seemed poised to reach truly unrivalled heights in 1989 when, in elections for the European Parliament, the UK Green Party received nearly 15% of the poll. The tendency to choose Green soon deserted the British voting public, but not before it had made a strong impact on the other mainstream parties who quickly began to emphasise their own environmental credentials.

Greening has been apparent elsewhere in society too. Advertisers have plundered the environment's comic resources, with puns about 'greenness' achieving wide currency and with jokes around environmental themes being employed to sell all manner of goods, from 'lid free' cars (convertibles) to 'nose-zone friendly' deodorants. Sports commentators have shown equal flexibility in weaving accounts of climate change into their remarks about unseasonably warm football matches as well as rain-affected cricket fixtures. Environmental thrillers have once again begun to populate television schedules and, for the late 1990s, Hollywood looks set to offer numerous environmental spectaculars, particularly on rainforest themes. Given all these trends, one can only assume that the public is increasingly familiar with certain aspects of environmentalism, and that such topics as acid rain, the ozone layer, and the greenhouse effect have attained daily currency. The penetration of these terms into everyday culture is, I suggest, indicated by a 1992 poster campaign in the UK for Regal cigarettes featuring a character (Reg) supposedly giving his views about topical issues, the joke (at least in many cases) being that he misconstrues his theme. One poster concerns his views on 'the greenhouse effect'. Under this heading Reg states simply that his tomatoes seem to grow better under glass. The advert only makes sense if the public at large is assumed to understand that the greenhouse effect 'really' refers to something else.

Many analytically significant questions arise from these shifts in public interest – for example, to do with the long-term impact of green consumerism or with the fate of ecological political parties. But, as the examples of global warming and ozone depletion indicate, one of the most striking features is the way in which widespread concern about environmental problems has brought a set of scientific issues to public attention. Until recently, very few people would have heard talk of ozone outside of the occasional chemistry class; now ozone features in jokes in television comedies and in popular newspaper stories about, for instance, Prince Charles' purported banning of his wife's ozone-destroying aerosols.

For this reason the study of knowledge and beliefs about the environment offers a key, topical example of the handling of science in the public realm. Of course, given the many different ways in which people can become interested in developing their knowledge about the environment, it would be wrong to assume that there is just one set of processes at work. Thus, some members of the public have become personally involved in environmental disputes, for example, over plans to mine gold in the mountains in the west of Northern Ireland or to site an incinerator in Derry;[2] in their case ecological politics have typically led to a pressing need to acquire familiarity with a specific range of scientific and technical arguments. In other cases the learning has been conducted in a much less urgent context, through general media coverage or through the environmental clamourings of people's children.

Some knowledge of environmental issues has become important in a wide range of occupations: from doctors alerted to the health effects of air pollution, through electrical appliance retailers (having to face up to questions about appliances' energy efficiency), to politicians who have generally wished to acquire enough expertise to avoid well-publicised blunders, such as President Reagan's celebrated assertion that trees are a major source of pollution.[3] Politicians' awareness of these topics has been reflected not only in their personal views, but also in the development of legislation and in the work of official bodies such as select committees reporting on sundry environmental issues. Lastly, environmental campaign groups have themselves been seeking to raise the public's awareness of the issues. These groups together have many hundreds of thousands of members just in the UK; their publicity material and stories routinely make press and broadcast-media headlines. They explicitly seek to influence public beliefs, and frequently publish 'briefings' on major issues such as the safety of underground storage of nuclear waste or the feasibility of wind power. And their views are backed up by a scientific warrant: the 'establishment' conservation bodies (for example the Royal Society for Nature Conservation) have long had a scientific ethos and large scientifically qualified staffs, but even the more radical groups now employ highly trained scientists on their campaign teams and commission original academic research. Indeed Greenpeace's official biographers boast that the organisation has equipped itself with the 'most sophisticated mobile laboratory in Europe'.[4]

These are just four of the routes by which public interest in the environment and public awareness of science are linked, and in this chapter I will only have the opportunity to examine one of them. My analysis will focus on the operation of environmental campaign groups and the way in which they offer to link scientific understanding to environmental concerns. According to one widely publicised interpretation, scientific investigation provides the facts upon which practical and policy decisions can be made – science, it is said, 'speaks truth to power'.[5] Particularly in the case of environmental policy, where it is assumed that nature cannot speak for itself, scientific representations of nature's needs might be seen as nature's stand-in. This study will look at the practicalities of trying to use science as nature's advocate.

Ambivalent responses to scientific authority

Anybody who hoped for a straightforward and harmonious relationship between scientific understanding of ecological issues and agreement on the practical steps to be taken will have been disappointed. One major cause of this disappointment has been the ambivalence of many environmentalists *vis-à-vis* the scientific enterprise. For one thing, many of our leading ecological problems can be seen as the *result* of our technological civilisation. For example, it was scientific research which first produced the pesticides which have posed such a hazard to wildlife and environmental quality since the 1960s.[6] Similarly, the CFCs which threaten the ozone layer are not naturally occurring chemicals; they were synthesised in scientific laboratories. In these ways, scientists can be seen to have been collaborators in much of the ecological destruction associated with our high-technology society. Many individual scientists are also closely associated with particular projects which are frequently viewed as environmentally damaging, such as nuclear power generation or intensive farming practices. There is at least a potential clash for greens who may experience a scepticism about the supposed benefits of scientific progress, but who are asked to accept that science is the key to successfully representing nature.

In the face of these problems, some environmentalists have been attracted to versions of the green argument which are principally founded on non-scientific forms of authority. For example, it is possible to seek to underpin an ecological worldview in conventionally religious or other spiritual ways. People can claim to gain a knowledge of nature's purposes and needs through this sort of inquiry.[7] But in secular Western societies these appeals can exercise only a limited attraction, and the principal form of legitimation in the leading environmental organisations remains that of scientific expertise. Even, as I have mentioned, such celebratedly anti-establishment organisations as Greenpeace draw increasingly on scientific authority, with Greenpeace (UK) having appointed an academic scientist as its director of science in London and Greenpeace International having established a laboratory at Queen Mary College, University of London.[8]

Of course, this emphasis on the scientific legitimations for green arguments is not to suggest that moral and ethical justifications are unimportant to environmentalism, particularly in relation to the non-human world. During the late 1980s the vocabulary of animal rights and animal welfare rapidly entered everyday language, indicating a fundamental change in common ways of considering animals, and signalling an expansion in the kinds of beings held to have moral rights.[9] But talk of rights has been most effective in promoting change in those cases where humans deliberately inflict suffering on individual animals, for example during hunts, in the production of furs, or in the course of animal experiments. Also, the vocabulary of 'rights' is customarily applied to individuals; rights of free speech, freedom of religious observance, and so on are attributed to individuals or individually to all members of a population. It is less clear how rights should be ascribed to abstract entities such as species or even the habitats which support species. Arguments about rights may prevent cruelty to particular seals, but they do not translate easily into firm guidance about the development of policies towards seal communities; can – for example – individual seals be killed without infringing the community's rights?[10] Furthermore, even if we take just those animals to which rights may well be granted, one still needs to know what it is that those animals want to use their rights for; once again, it is typically the job of scientific advocates to inform us what animals need and want.

The common ambivalence about the authority to be accorded to science may even tend towards antipathy among groups which are ideologically opposed to features of our current technological civilisation and among animal welfare activists who find that research scientists number among their campaign targets. In these cases, green activists and supporters have often come to have a distrust of experts and of scientific pronouncements. They are aware that reputable scientists have supported policies to which these groups are opposed (such as nuclear power generation), and also that governments and established interests often use demands for scientific proof as a way of delaying practical action. In the UK the most conspicuous example of this strategy was its use by the CEGB and – initially at least – government spokespersons as well, to argue that there was no certain proof that acid emissions from UK power stations were responsible for the acidification of lakes and the death of trees in Continental Europe. Official agencies could protest that they were as concerned as anyone but, until there was conclusive evidence that it was their power stations' waste gases which caused the pollution, it would be irresponsible to spend taxpayers' money on reducing emissions.[11]

Once ambivalence shades into opposition, green groups face a serious difficulty. They have good grounds for distrusting scientific authority, but have no other place to turn for universalistic, definitive answers. Their own occasional impatience with scientific procedures, arising from a desire to take prompt practical action and from a distrust of the motives behind delays in arriving at officially recognised scientific conclusions, opens them to attack from outside observers on the grounds that they lack objectivity.[12]

Given that environmental campaigners themselves have a complex and ambivalent attitude to scientific authority and the canons of scientific argument, it would be unrealistic to suppose that the public presentations of the relationship between science and ecological issues would not reflect these tensions. Environmentalists, offering themselves as nature's advocates, contribute a mixed message to the public's understanding of science.

Indeed, so central is this tension within the greens' case that Beck has nominated environmental protest as a prime example of the crisis he diagnoses in the modernist worldview, a crisis he terms 'reflexive modernisation'. In general terms this crisis arises when modernist principles are destructively applied to themselves (that is, reflexively). Beck chooses the environmental example because it displays the problems arising from the application of science to itself. Thus, for Beck, the systematic application of critical analysis to scientific practice and the philosophy of science has identified limitations in science's cognitive authority. In the acid rain case, for example, scientific effort was invested in finding the cause of acid rain; at the same time other scientists worked to demonstrate that science could not prove, beyond reasonable doubt, that Scandinavian acid rain arose from Britain. As he expresses it:

> science is involved in the origin and deepening of risk situations in civilization and a corresponding threefold crisis consciousness. Not only does the industrial utilization of scientific results create problems; science also provides the means – the categories and the cognitive equipment – required to recognize and present the problems *as* problems at all, or just not to do so. Finally, science also provides the prerequisites for 'overcoming' the threats for which it is responsible itself.[13]

There are two noteworthy features in Beck's formulation of this issue. First, he presents the tension or 'paradox' as a problem for the legitimacy of science; at the same time, however, as I have just argued, it is a problem for the legitimacy of environmental campaigners' views. And, at least to date, given the imbalance of power and authority between campaigners and official bodies, the difficulty has been more acute for environmentalists wishing to change the *status quo* than for the authorities and the establishment scientists seeking to maintain it. Second, his account of the paradox presents it in a rather idealistic light. In other words, it appears that science is driven to this self-critical position by the inescapable force of reason alone.[14] However, an examination of the practical business of putting science to work in environmental organisations will allow us to appreciate the significance of social context in generating this tension and in shaping the way it is handled.

Contrasting strategies for science as nature's advocate

In the light of these tensions and difficulties, it is only to be expected that the treatment of scientific information and scientific authority would vary from one environmental organisation to another. A large part of this variation can be understood by referring to a distinction, made by some commentators, between 'environmental' groups and 'conservation' organisations; typically, according to this view, the former (such as Friends of the Earth (FoE)) work through campaigning to 'confront the negative effects on the environment of late-twentieth century society and [to] try to ameliorate them'.[15] For their part, the latter (including the Royal Society for Nature Conservation (RSNC)) aim to conserve and enhance existing habitats and species.

It is important not to exaggerate this dichotomy. The Royal Society for the Protection of Birds (RSPB), a clear candidate for the second group, could none the less justly be said to have been established to confront the 'negative effects on the environment' of late nineteenth-century society.[16] In any case, organisations such as the RSNC and the RSPB, as well as smaller bodies including Plantlife and the Marine Conservation Society, are increasingly adopting a policy of campaigning.[17] In large part this is because they recognise that it is often more cost-effective to conserve land or habitats through lobbying for alterations in, for example, planning procedures or the Common Agricultural Policy than by trying to raise money to buy reserves – particularly since isolated reserves will be of little use to large mammals or many birds, which feed over a large area and thus depend on the environmental quality of surrounding farmland.

However, the dichotomy is revealing in one way: conservation groups typically have a background in natural history, and their membership has generally been dominated by scientists and amateur naturalists. As a result their ethos was different from that of the environmental groups which started off in the late 1960s and early 1970s as groups critical of contemporary Western society.[18] This difference in background and ethos has significant implications for these groups' approach to science and scientific authority. A brief introduction to the RSNC will illustrate the part played by science in its ethos and practice.

Founded with an elite scientific membership just before the First World War as a Society for the Promotion of Nature Reserves (SPNR), the Society's initial objective was to encourage the protection of sites of importance for natural history. Such reserves were for 'the enjoyment of lovers of wild nature, the pursuit of scientific knowledge, and the well-being of the community in general'; where these objectives threatened to clash, the Society favoured scientific priorities.[19] The Society's influence grew, particularly because in the 1940s the government invited it to assist in developing a strategy for national nature conservation; accordingly, in Lowe's words, in mid-century the 'ecologists gradually assumed the leadership of the conservation movement'.[20] This position was consolidated in 1949 with the formal establishment of the Nature Conservancy (the 'first official *science-based* environmental conservation agency in the world'[21]), the membership of whose leading council overlapped strongly with that of the SPNR. The Society and the various official nature conservation bodies have remained close ever since, in particular sharing an emphasis on the use of scientific criteria in assessing conservation merit. The flavour of legislation has also been decisively shaped along these lines, with the most ubiquitous conservation designation in the UK being the Site of Special *Scientific* Interest (SSSI).

The Society also experienced a second significant institutional development. In the post-war period, through to the 1960s, there was a rapid growth in the number of county Wildlife Trusts, regional bodies associated with the Society, but chiefly concerned with local nature conservation, reserve acquisition, and site management. The SPNR became the national co-ordinating body for these groups and accordingly changed its name (by a two-stage process in 1977 and 1981), to the RSNC. These Trusts took on much of the ethos of the central body being dominated in the early stages by natural historians, scientists, and enthusiasts. Although not as closely bound to the science of ecology as the RSNC and the Nature Conservancy (later the Nature Conservancy Council (NCC)), the Trusts retained a scientific ethos. Initially run by volunteers, they were dominated by people with a scientific understanding of, and interest in, wildlife, a characteristic which lent the Trusts important strengths, but also weaknesses.

The practicalities of a scientific ethos

The benefits which organisations derive from a scientific ethos are essentially twofold. First, their scientific standing, supplied by the academic scientists and civil servants who serve as volunteer members and – increasingly – by scientifically trained staff, has enabled the RSNC and the Trusts to deal in an authoritative manner with the Department of the Environment (DoE), the NCC, and local authorities. They have shared many of the same objectives and concerns; staff have moved from one sector to the other; and the DoE and NCC have been happy to let the Trusts engage in reserve management, and have even leased reserves to them. Even when there has been disagreement (for example over the DoE's willingness to sacrifice certain SSSIs to developers), the official bodies have had to recognise the expertise of the voluntary groups; it has not been possible to dismiss their arguments as scientifically ill-informed. Overall, organisations such as the RSNC and RSPB have enjoyed good relations with official conservation agencies, underwritten by respect and friendship between their respective staffs. As Tim Cordy, chief executive of the RSNC, acknowledged, these organisations tread a fine line: 'It is crucial that the voluntary sector works *with* government, but not *for* it.'[22]

The character of the voluntary conservation bodies has also yielded financial benefits. For one thing, since they undertake projects similar to those which official agencies have to carry out, and since they tend to do them relatively cheaply, the organisations have been eligible for a good deal of grant aid. In the case of some Trusts and various 'Link' organisations (which co-ordinate conservation bodies' work) government grants of various sorts constitute the great majority of their income; indeed, so great was the apparent funding for the County Trust for Northern Ireland (the Ulster Wildlife Trust (UWT)) that a speaker from the Northern Ireland DoE joked at a UWT meeting in 1988 that the Trust would soon be a greater drain on government finances than the then state-owned and ailing shipyard.[23]

Good scientific standing also offers a further avenue for income generation through the sale of expert advice. Trusts affiliated to the RSNC have been among the leaders in developing consultancy services, whether providing environmentally sensitive landscaping advice to schools and hospitals or preparing environmental statements for developers. Environmental consultancy is a doubly important area of work for green groups. It earns them good, 'free' money (not tied to government departments) and, second, the Trusts and similar groups believe that, if they perform the work themselves, this will maximise benefits to the environment since they view their own consultancy advice as single-mindedly wildlife friendly.

The practicalities do not, though, all work in the same, positive direction. The scientific ethos of conservation organisations may result in tasks which demand scientific expertise (surveying, field observation, reserve management) taking precedence. Scientifically trained and oriented staff are likely to feel happier doing this kind of work than campaigning or lobbying, even though, as noted above, in many instances campaigning may be a more cost-effective approach to nature conservation than direct intervention (that is, managing or acquiring sites, a labour-intensive undertaking). In conservation groups, in which chairpersons, many directors, and some development officers have come from a scientific background, there will not tend to be a ready familiarity with managerial, fund-raising, and publicity skills. Moreover, because friendship networks typically revolve around people's occupations and interests, predominantly scientific groups may have few acquaintances in management or commerce on whose help they can call.

This disjunction between a primarily scientific and an organisational outlook can be especially acute in relation to the setting of campaign priorities. Natural historians will typically have their own conservation priorities, based on rarity or on scientific interest. These priorities will not necessarily coincide with the issues best calculated to excite public interest and media attention; put bluntly, while a drab plant may be endangered or be botanically exhilarating, majestic colourful birds offer the best 'photo-opportunities'. The tensions involved in attempting to meet these conflicting demands were exemplified during discussions among members of the UWT's scientific committee; in considering which issues to highlight in promotional material, they moved on from considering suggestions for badger week and otter campaigns to muse on the possibility of slug week and rat year. These jokes are a symptom of an anxiety that promotional needs – the organisational requirement to find a popular, newsworthy campaign target – will be elevated above scientific priorities. Similar concerns may arise in the case of publicity material too. When the RSNC unveiled its revised badger logo, some naturalists were more concerned with the uncertain sexual identity of the pictured badger than with its likely public-relations value. Such concerns are further attested to by naturalists' humour, in this case a proposal that the UWT's slogan should be: 'protecting your local biotopes'.

Such tensions between different aspects of these organisations with contrasting objectives are likely to be intensified by current trends in the environmental movement. Increasingly, environmental organisations are having to compete with each other to gain public attention and support, and for this reason it is important to be associated with leading, newsworthy issues. The background to this competition is quite straightforward. The pressure groups are not quite like businesses competing for market share: they co-operate a good deal, and members are not exactly like customers since they may well subscribe to several groups. Of course, ordinary members will limit their subscriptions at some point, but the real competition is for major sponsors – increasingly, companies, but also charitable trusts – who only have specified budgets to disburse. Under these competitive conditions, fund-raising and publicity success is to some degree self-perpetuating: firms will fund campaigns which have a high profile, a profile further heightened by this backing. No groups can afford to miss out on these important market opportunities. Moreover, such competition operates to concentrate more attention on the highest-profile issues (such as the rainforests, the conservation of attractive species, and so on) and can leave other environmental issues 'orphaned'.

Both the RSPB and the RSNC have responded to the challenge: the former by following birds along migration routes and away into the rainforests, thus associating their organisation with high-profile, international environmental issues, and the latter by identifying itself with cherished members of Britain's wildlife and with countryside values.

In summary, the conservation groups are increasingly having to follow a commercial and market-oriented logic which leads to different policies and priorities than would follow from a narrowly scientific interest in conservation. Thus, they are being led away from a 'purely' scientific vision of the environmentalist's project not, as

Beck might have anticipated, by the corrosive rationality of reflexive modernity, but by the perceived demands of competitive pressure-group politics. These recent trends have had a large impact on scientifically dominated nature conservation groups.

Environmentalism without a scientific ancestry

In essence, 'environmental' organisations lack both the weaknesses and the strengths outlined above. Thus, even had they wished to, they could not have enjoyed such close, practical co-operation with official agencies as has the RSNC. Nor have they been able to earn their green pounds in the same ways. They have, however, been more able to respond to certain market forces, particularly the 'markets' of public opinion and of media coverage; Greenpeace in particular has been very astute in managing news coverage of its campaigns. Crucially for my argument, these organisations have had more freedom to be critical of scientific opinion and expert judgement. They have often given voice to the idea that scientific evidence is manipulated by official agencies, and that government-commissioned scientists have asked the wrong sorts of questions. In their report to mark the twenty-first birthday of British Nuclear Fuels, FoE cite, with heavy irony, the BNFL chairman's claim that 'Technically, we know how to deal with the waste'; the report then goes on to list doubts, misgivings, and accidents which underline the ironic tone.[24] The supposedly scientific and technical statements of officialdom are seen as commonly containing a political agenda, and as a reflection of how the authorities wish an issue to be viewed.

That they have felt able to criticise the scientific establishment does not imply that environmental groups have been free of campaigning constraints; Friends of the Earth (FoE), for example, has been studied in its avoidance of party-political partisanship. But, since scientific study of the natural world is of less inherent appeal to their supporters, since the organisation's work depends more on campaigning than on surveying and management, and since these groups refuse to engage in consultancy work, they can retain some distance from science.

However, in its turn, this distance may result in difficulties. In some cases, the information necessary for campaigning can be straightforwardly derived. Thus in the celebrated case of (UK) FoE's popular and successful campaign against CFC-driven aerosols (in 1988–9) it was enough to argue that, if American firms could withdraw CFCs from deodorants and other spray-can products, then so could British ones.[25] An identical argument was used in the late 1980s by Greenpeace against Ford's failure to fit catalytic convertors or other emission control systems on its British models. Borrowing the company's own slogan ('Ford gives you more'), the Greenpeace campaign pointed out that a Ford car in Britain gives vastly more toxic pollution than one bought in the USA because of the company's response to the different pollution control regulations.[26] Essentially in both cases the argument is about double standards. If something is technically possible in one developed country, it must be equally possible in Britain. The argument does not need to appeal to any external indicators of correctness.

When it comes to contested scientific information, these groups are in a less comfortable situation. Thus, on 12 August 1990 a programme was broadcast on Channel 4 in the UK casting doubt on the reality of global warming as a result of the greenhouse effect. It set out to question belief in warming, and implied that climate scientists might be led into making exaggerated claims about climate change because the existence of such a threat would make it easy to acquire the resources for their research programmes.[27] In the FoE *Local Groups Newsletter* a staff member attacked the broadcast by appealing to the weight of majority scientific opinion. The testimony of 'around a dozen dissident scientists' had to be weighed against the views of the 300 or so scientists who 'wrote and peer reviewed' the report to the Intergovernmental Panel on Climate Change (IPCC).[28] The article went on to bolster this argument by rhetorically pointing out that the IPCC scientists could hardly be portrayed: 'as a raving bunch of eco-anarchist nutcases. Some would say they are a cautious bunch, wary of hard-won scientific reputations and not prone to wild exaggeration'.[29] The artful use of the expression 'some would say' seems to imply that this opinion is universally shared, rather than just being the view of FoE.

No doubt it is quite reasonable for FoE's supporters to argue in this way. But the point arising from this example is that FoE has no scientific evidence of its own to use in settling this controversy. FoE's best argument is just that the most well-informed scientific opinion is on their side. They find themselves invoking the 'consensus of scientific opinion' to overcome the TV journalists' deconstructive arguments – exactly the opposite of strategies environmentalists have themselves deployed in other contexts (for example, against the official consensus on the supposed safety of nuclear installations).

We have already seen how those organisations with a scientific ethos have been altered by the growing demands of competition. We can anticipate a corresponding change in this second type of environmental group which is likely to be affected by growing scientific and technical competence. As technical expertise increases, confidence grows in one's ability to win arguments through persuasion, and this may lead groups to adopt a less strident campaign style; in his study of the British green movement, McCormick notes that through the 1980s Greenpeace became 'less confrontational, and more inclined to use the same tactics of lobbying and discreet political influence once reserved by the more conservative groups'.[30] But campaign groups have, so to speak, already lost their innocence about the neutrality of 'pure' scientific reasoning; they have, in Beck's terms, already experienced this aspect of the 'crisis' of late modernity.

Given their familiarity with the shortcomings and ambiguity of scientific evidence, and given their distrust of parts of the scientific establishment and of the authorities' use of science, they cannot be expected to embrace the path of campaigning through expert argument. But this means that there is no single approach for them to adopt. Admittedly, they can appeal to certain novel methodological principles, such as the 'precautionary principle': the principle which stipulates that new substances or new processes should only be used once they have been shown to be harmless. But even then there will be room for negotiation and controversy about the 'proof' that, say, substance X is harmless. Accordingly, the likeliest outcome is that campaign organisations will adopt a flexible approach, a mix of academic science and pragmatism, informed by their experience of what makes for campaign success. As a graphic example, it seems extremely unlikely, whatever marine scientists found out about whale populations or whatever claims came to be made about whales' lack of intelligence, that campaigners' opposition to whaling would decline.

Such pragmatic epistemological flexibility is, in any case, further encouraged by practical limitations on the use of scientific knowledge in the service of campaigning. For one thing, even groups which have a large scientific staff or can count on assistance from sympathetic academic researchers find that they cannot gain access to all the information they would like. They do not have the budgets to subscribe to all the publications they might desire, nor can they maintain extensive libraries. In any case, academic science – even in ecology – will not necessarily generate the kinds of research they would like to see done. One final dilemma confronts the large organisations which conceivably would have the resources to undertake some research. For example, the RSPB has obligations to spend money on its reserves and on practical bird conservation, while Greenpeace needs to fund its campaigns and its ships. Against such practical and pressing expenditures, a research budget is hard to justify.

In consequence, such organisations are dependent on scientific knowledge produced by other persons or agencies, knowledge which is suited to the objectives and agenda of those other groups. Even if it were attractive for environmental groups to try to take on the mantle of green science, they would be confronted by severe practical restrictions.[31]

Conditioning the way science is put to work

For green organisations, it is clearly a major practical challenge to work out how to treat scientific knowledge claims and how to deal with scientific expertise. In their various ways, depending on their ethos, ancestry, and current mission, these groups have to both use and criticise scientific logic, the paradigm of cognitive modernity. We have examined several of the tensions and paradoxes to which this gives rise.

However, in most of the instances analysed so far it has more or less been left to the organisations themselves to figure out their response to these tensions. Thus, Wildlife Trusts can make their own judgements about the scientific value of their nature reserves; FoE can formulate its own stance on the conflicting evidence about global warming.

However, in certain significant institutional contexts, these groups have much less control over the handling of, say, scientific evidence on the emission of acidic gases or on threats to indigenous wildlife. Two institutions are of particular importance, both in terms of their public prominence and because of the way they have focused attention on the tensions inherent in the treatment of scientific expertise. They are the media and the law.

The media

Turning first to the media, it is clear that they frequently feature environmental news. The appeal of such news is described by Lowe and Morrison, who point out that:

> At the editorial level, a major attraction of environmental issues is that they are public interest issues of a non-partisan nature. Thus they provide an important outlet for campaigning and investigative journalism even for newspapers which take a typically conservative stance on other matters and for broadcasting services striving for a 'balanced' view.[32]

Media interest in the environment has made a great contribution to the recent upsurge of public concern about green issues. But the media have their own demands, preferring certain themes and stories over others. These preferences have acted to shape the public face of environmental campaigning. For example, media interest tends to reinforce the accent on the photogenic and the picturesque.[33] Some environmental groups have responded better than others to the demands for media-friendly campaigns. Typically also, the mass media will seek comment on an issue when programme makers perceive that issue as newsworthy, rather than at the point when green groups have the technical information prepared to their own satisfaction. This fact has had an impact on environmentalists' attitude to their scientific preparation, sometimes leading campaigners to calculate how little knowledge they can 'get away with', rather than seeking to be as exhaustively knowledgeable as possible.

Media conventions can also influence how debates are handled and develop. Thus, the documentary and news formats encourage a presentation of environmental issues as a balanced debate between two 'sides': for example, for and against nuclear power generation or for and against flue gas desulphurisation. In this way, competing views which insiders regard as very unequally matched may be accorded similar levels of respect. Media treatment can occasionally have very far-reaching consequences for a controversy, as was the case with a plant-growth regulator known as Alar. The substance's manufacturer and the US Environmental Protection Agency (EPA) were engaged in a protracted disagreement about exactly how (un)safe this chemical was, when a 'public interest science' group, the Natural Resources Defense Council (NRDC) released its own results, indicating that the danger was hundreds of times greater than even the EPA had proposed. These findings were publicised on a nation-wide news broadcast, resulting in a huge public outcry: 'Within months, [the manufacturer] concluded that it would no longer be profitable to keep Alar on the market and announced that it would voluntarily withdraw the product.'[34] In this case, media publicity overrode the processes of technical debate, ensuring that only the NRDC's assessments of risk counted.

Although this case is an extreme one, it illustrates my point that agencies external to green groups can decisively shape the public presentation and credibility of their scientific arguments. Accordingly, environmentalists fashion their treatment of scientific and technical considerations, not only in the light of the perceived strengths and weaknesses of scientific styles of reasoning, but also with regard to the way that influential external agencies are likely to handle these matters.

The law

The way in which legal institutions condition the use of science varies from country to country. In the USA, where official actions are routinely open to legal review, the law's influence has been very great; in Britain its impact has been less pronounced. As Jasanoff has demonstrated, US legal review has affected not only the substance of environmental policy, but also the way that scientific arguments are deployed in making and defending decisions.[35]

In essence, the adversarial legal process is geared towards the deconstruction of people's arguments; scientific judgements, as much as any other, have fallen prey to this deconstruction. Since scientific decisions do not follow a mechanical, routine method, but depend on the skilled exercise of expert judgement, legal examiners have repeatedly found it possible to find conflicting opinions about any controversial scientific matter (whether nuclear power plants are safe, whether pesticides cause cancer, and so on); they have also been able to show that individual experts' opinions depend on judgements which cannot be defended by appeal to straightforward and transcendental principles. In other words, it has been possible to make scientific views appear like 'mere' opinion.[36] In this sense, the adversarial legal process has been able to practise (destructive) reflexive modernisation on science (even if science has not done it to itself).

The responses to this fact have been varied. In the USA the EPA has adopted what may appear to be an ironic strategy: 'Under continual assault from political adversaries, EPA's environmental science has more and more justified itself in terms of its legal, institutional, and procedural underpinnings rather than the truth-value of the facts it alleges.'[37] Jasanoff also notes a repeated tendency for the authorities to try to separate out the 'science' from the 'policy' elements in environmental judgements. However this undertaking was constantly undermined by the acknowledgement that the two things were inseparable: thus, apparently factual procedures for recognising whether a substance was liable to induce cancer were influenced by assessments of how much potential for error was acceptable. The fusion of scientific and policy principles was indicated by the fact that EPA presented the same principles to different administrative audiences as scientific fact, on one occasion, and as policy proposals, on the other.[38] These legal examinations of scientific arguments about the environment had a double impact. First, they served to throw doubt on the credibility of specific scientific claims. Second, they revealed in a very public forum the inherently negotiated character of the 'proper' boundaries of science, indicating that the very identity of science itself is socially constructed.

Given the more informal regulatory procedures prevalent in the UK, these issues have not come so clearly to light. However, scientific evidence offered by environmentalists has come to grief at the hands of solicitors' deconstructive arguments. Moreover, conservation groups' attempts to derive income from consultancy work have also suffered embarrassment at the hands of solicitors when their clients' development proposals have gone to court. That the consultants have a presumptive interest in conservation (due to their institutional affiliation) has been exploited by lawyers to suggest that the expert testimony is not impartial, but, in fact, slanted towards their groups' campaign interests. In one such case, the Ulster Wildlife Trust's representatives were angered that the impartiality of their testimony was called into question; they insisted that they had carried out a purely scientific survey. Here again, the scientific ethos of the organisation has left it exposed to external criticism.[39]

Legal examination has therefore had a large impact on certain specific environmental judgements. But the nature of adversarial cross-questioning has also led to the public exposure of apparent weaknesses in the scientific basis of environmental arguments, and has called into question the exact boundaries of science itself. Lawyers have refined and consolidated their methods for challenging scientific evidence, whether it derives from official agencies, such as the EPA, from industry, or from campaign organisations. For this reason, the legal advocates have been even more influential than the media in exposing the tensions in, and weaknesses of, science as nature's metaphorical advocate.

Conclusion

In modern Western societies it is accepted that wildlife and the natural environment need advocates; scientific expertise has emerged as the form of advocacy which commands the greatest legitimacy. A scientific interest in nature also motivates many supporters of conservation organisations, while the organisations themselves harness scientific expertise for practical tasks such as reserve management and the monitoring of biodiversity. But science's advocacy role is far from straightforward.

For one thing, many supporters of environmentalism have misgivings about our scientific civilisation and about the social benefits which supposedly derive from science and technology. Problems have also arisen with voluntary organisations' deployment of science: in the case of nature conservation organisations a dominant scientific ethos has tended to lead to managerial and administrative difficulties while, for campaigning groups, increasing reliance on scientific expertise has exposed a potential conflict with their scepticism about official science.

In these ways environmental groups have experienced something akin to the reflexive modernity of which Beck wrote. These challenges to the practical sufficiency and to the universal validity of science have been amplified by the workings of two external social institutions: the media and the law. Both these institutions have acted in a twofold way on the role played by science in environmental debates. First, they have both offered challenges to the way ecological evidence is interpreted; thus, for example, the law has repeatedly been used to question the validity of the EPA's rulings on risk, while the media have been able to focus public and political attention on sensational – but possibly minority – views. Second, both institutions have called into doubt the exact boundaries of scientists' competence. Jasanoff showed how, under legal examination, the EPA drew the boundaries between science and policy in a variety of ways. Media attention too has questioned and blurred the boundaries around legitimate science.

In conclusion, therefore, we can see that the experiences of environmental and conservation organisations highlight many of the problematic aspects of attempting to employ scientific expertise in areas of public concern. These groups have tried to promote the public understanding of environmental (and thus often scientific) issues and to make the public more aware of the arguments about the value and trustworthiness of expertise. From the practical experiences of environmental organisations reviewed in this chapter, it appears that there is no single, simple way of harnessing scientific expertise to the public interest; groups need to strike a pragmatic balance between accepting and denying the overriding validity of science. Moreover, the systematic and legal-rational examination of scientific evidence and judgement, particularly when conducted in public fora, is increasingly leading to the weakening of science's practical authority and of its claims to transcendental validity.

In this sense, my conclusion is a post-modern one: environmental controversies have encouraged the development of reflexive modernity, an attitude which fits with environmental campaigners' pragmatism. But we can also see that the way this process has taken shape depends on the character of the predominant environmental groups in any country, on its legal and administrative system, and on its media. The practical utility of science and the public perception of its validity and adequacy are not determined by any logic of post-modernity, but by the practical decisions made by campaign organisations, and by the procedures adopted by the media and by nations' legal authorities.

NOTES

1. The research reported in this chapter was supported by an award (A0925 0006) from the ESRC and Science Policy Support Group under the Public Understanding of Science Initiative. I should like to thank all those members of environmental groups and other respondents who gave time to be interviewed and observed during my research work, a two-year participant observation and interview study of non-governmental environmental groups in Northern Ireland. The study was supplemented with analyses of the corresponding groups in Great Britain and with published analyses of environmental politics in other countries.

2. See respectively, Allen R., and Jones, T., *Guests of the Nation* (London: Earthscan, 1990) p. 78 and Allen, R., *Waste Not, Want Not* (London: Earthscan, 1992), pp. 3–37.

3. See Green M., and MacColl, G., *There He Goes Again: Ronald Reagan's reign of error* (New York: Pantheon, 1983) p. 99, where the then President is quoted, from the journal *Sierra*, stating that 'Approximately 80 per cent of our air pollution stems from hydrocarbons released by vegetation.'

4. Brown M., and May, J., *The Greenpeace Story* (London: Dorling Kindersley, 1989), p. 150.

5. On this metaphor see Jasanoff, S., 'Science, politics, and the renegotiation of expertise at EPA', *Osiris* 7 (1992) 1–23, on p. 2.

6. Nicholson, M., *The New Environmental Age* (Cambridge University Press, 1987), pp. 46–51.

7. See the discussion in Spretnak, C., and Capra, F., *Green Politics: the Global Promise* (London: Paladin, 1985) pp. 230–58.

8. This ambivalence is examined further in Yearley, S., 'Green ambivalence about science: legal-rational authority and the scientific legitimation of a social movement', *British Journal of Sociology* 43 (1992) pp. 511–32, where Greenpeace UK's appointment is also discussed; on the QMC laboratory see Pearce, F., *Green Warriors: the people and the politics behind the environmental revolution* (London: Bodley Head, 1991), p. 39.

9. See Dobson, A., *Green Political Thought* (London: Unwin Hyman, 1990) p. 38, and Warren, M. A., 'The rights of the non-human world', *Environmental Philosophy*, edited by R. Elliot and A. Gare (Milton Keynes: Open University Press) pp. 109–34.

10. See Wenzel, G., *Animal Rights, Human Rights* (University of Toronto Press, 1991).

11. The unfolding of this argument is discussed in A. Irwin, 'Acid pollution and public policy: the changing climate of environmental decision-making', *Atmospheric Acidity: sources, consequences and abatement*, edited by M. Radojevic and R. Harrison (Amsterdam: Elsevier, 1992) pp. 549–76.

12. See Yearley, 'Green ambivalence' and, as an example, North, R., 'Greenpeace: still credible?', *The Independent*, 21 September 1987, p. 15.

13. Beck, U., *Risk Society: towards a new modernity* (London: Sage, 1992) pp. 156 and 163.

14. This point is also made by Scott Lash and Brian Wynne in their introduction to Beck's volume, where they suggest that it is (unwise but) possible to argue that 'the religion of

science secularizes itself, is pushed through the barriers of its own precommitments by the impetus of criticism built into [its] social structure', p. 6.

15. Dobson, *Green Political Thought*, p. 3.

16. The RSPB, which celebrated its centenary in 1989, originated as a campaigning organisation opposed to the heedless killing of birds for the feather trade. By the middle of this century it had turned into the body for bird enthusiasts which essentially it still remains, although see below. On its history see Samstag, T., *For Love of Birds* (Sandy, Bedfordshire: RSPB, 1988), and Yearley, S., *The Green Case: a sociology of environmental arguments, issues and politics* (London: Routledge, 1992) pp. 61–7.

17. Yearley, *The Green Case*, pp. 56–61.

18. For all practical purposes Greenpeace was founded in 1971 and Friends of the Earth in 1970; see Yearley, *The Green Case*, pp. 67–9, and Lowe, P., and Goyder, J., *Environmental Groups in Politics* (London: Allen and Unwin, 1983) pp. 124–37. See also Beck, *Risk Society*, p. 162.

19. The words of the SPNR cited in Sheail, J., *Nature in Trust: the history of nature conservation in Britain* (Glasgow: Blackie, 1976) p. 62; on the pre-eminence of scientific considerations see Lowe, P., 'Values and institutions in the history of British nature conservation', *Conservation in Perspective*, edited by A. Warren and F. B. Goldsmith (Chichester: John Wiley, 1983) pp. 329–52, especially p. 341.

20. Lowe, 'Values and institutions', p. 342.

21. Nicholson, *The New Environmental Age*, p. 95, emphasis added.

22. Cordy, T., 'A manifesto of commitment', in the RSNC's magazine *Natural World* (Autumn 1991), 5. The organisation is now known as 'The Wildlife Trust'.

23. For more on the Ulster Wildlife Trust see Yearley, S., and Milton, K., 'Environmentalism and direct rule: the politics and ethos of conservation and environmental groups in Northern Ireland', *Built Environment* 16 (1990) 192–202, especially 195.

24. Friends of the Earth, *British Nuclear Fools plc* (London: FoE Ltd., 1992), p. 9.

25. Friends of the Earth, *The Aerosol Connection* (London: FoE Ltd., 1989).

26. *The Guardian*, 27 October 1988, p. 5.

27. Channel 4, *The Greenhouse Conspiracy* (London: Channel Four Television, 1990) p. 27.

28. Dilworth, A., 'Global warming', *FoE Local Groups Newsletter*, no. 186, September 1990, p. 19.

29. Ibid.

30. See McCormick, J., *British Politics and the Environment* (London: Earthscan, 1991) p. 158.

31. For more on these practical limitations see Yearley, 'Green ambivalence', and Cramer, J., *Mission-Orientation in Ecology: the case of Dutch fresh-water ecology* (Amsterdam: Rodopi, 1987) pp. 49–62.

32. Lowe, P., and Morrison, D., 'Bad news or good news: environmental politics and the mass media', *Sociological Review* 32 (1984) 75–90, on 80; see also Lowe, P., and Flynn, A., 'Environmental politics and policy in the 1980s', *The Political Geography of Contemporary Britain*, edited by J. Moran (London: Macmillan, 1989) pp. 255–79, especially p. 269.

33. Lowe and Morrison, 'Bad news', 81.

34. Jasanoff, S., 'American exceptionalism and the political acknowledgment of risk', *Daedalus* 119 (1990) 61–81, on 74.

35. Ibid., and also her 'Science, politics, and the renegotiation of expertise at EPA', *Osiris* 7 (1992) 1–23, and 'Cross-national differences in policy implementation', *Evaluation Review* 15 (1991) 103–19.

36. See Yearley, S., 'Bog standards: science and conservation at a public inquiry', *Social Studies of Science*, 19 (1989) 421–38.

37. Jasanoff, 'Science, politics, and the renegotiation of expertise at EPA', 3.

38. Ibid., 13.

39. This case is reported in Yearley, S., 'Skills, deals and impartiality: the sale of environmental consultancy skills and public perceptions of scientific neutrality', *Social Studies of Science* 22 (1992) 435–53. See also Yearley, S., 'The environmental challenge to science studies', in *Handbook of Science and Technology Studies* edited by S. Jasanoff, G. E. Markle, J. C. Petersen, and T. Pinch. (London and Beverly Hills: 1995), pp. 457–79.

9 Proteins, plants, and currents: rediscovering science in Britain

HARRY ROTHMAN, PETER GLASNER,
AND CAMERON ADAMS

Introduction

As discussed in previous chapters, one important dimension of the contemporary public understanding of science is the heterogeneity of scientific knowledges and understandings. Whilst conventional perspectives portray science as a unitary and coherent body of knowledge and expertise, the accounts in this book have typically stressed the *diverse* character of science as it is encountered by various publics. In looking to the future development of science–public relations it is important also that we consider the *changes* which are currently affecting modern science – not least because these will help form the new context for these relations. In a book devoted to the linkages between science and the public we therefore need to examine the various and shifting understandings of science within the scientific community.

Accordingly, this chapter seeks to identify how science is contextualised by academic scientists, administrators, industrialists, and members of some environmental organisations, and distinguishes between the various, and largely disparate, 'understandings' of strategic science held and enacted in the civil sector of British science. It is based on an investigation conducted in the context of policy changes at the political level which have borne most directly in recent years on the development of science and technology policy in Britain. There are several policy changes which deserve special consideration with regard to British civil science in the 1990s. First, the move towards the *prioritisation* of key areas of scientific research using special programmes for the focusing of government funding. Second, the fact that these have evolved within an increasingly stringent *economic framework*. Third, the *narrowing of the institutional base* for this work through a process of rewarding 'excellence' and the concentration of resources in key institutions. Furthermore, the British Government has sought to encourage a greater proportion of external non-government funding in all areas of scientific endeavour, and has of necessity encouraged a greater emphasis on shorter-term collaborative research. Finally, the promotion of the government's political and economic goals causing a concomitant shift away from pure research towards *more strategic areas.* (Ince 1986).

The principal findings of this chapter suggest that people situated at the various institutional levels concerned with strategic science promote differing, and largely disparate, 'models' of science in terms of its theoretical basis, its development, and its application. This contrasts sharply with the bulk of discussion of the public understanding of science which assumes a homogeneity of views within those communities we studied. Our research also indicates an increasing prominence in the scientific community of groups and individuals whose primary interests are instrumental and strategic. Again, this contrasts with the orthodox view which emphasises curiosity-driven research. Finally, our results attest to the marked tendency for strategic research to be subjected to the requirements of commercial confidentiality. All this suggests certain major characteristics of contemporary science which will in turn affect the possibility for new relations between science and public groups.

The 'Save British Science' (Connerade 1988) campaign has highlighted the shrinking nature of the science base in the UK over the last ten to fifteen years, providing the broader political context within which our study developed. This 'disappearance' of funding for science from government has been an essential element in government strategy for concentration upon areas of excellence and the transfer of responsibility for some research funding to non-governmental sources. We see this as 'disappearing science' at the 'political' level. Although we do not investigate defence or military science, investigation is, by definition, undertaken discreetly and is thereby hidden from public view, even though spending on military R and D consumes the largest proportion of available funding in the UK. This could be counted as another area in which science 'disappears'.

However, our research has found that science can be described as 'disappearing' on at least two other levels: the 'conceptual' and the 'commercial'. We discovered that, even within the scientific community, actors hold a number of different, and often conflicting, views about what constitutes scientific activities. Science can thus be said to be 'disappearing' in that the conventional view of the work of the scientist represents only one of a number of different understandings, which the public at large rarely identify with scientific activity. A shift in the balance away from unfettered blue-sky or basic research into strategic and more applied areas thus gives the illusion of a contraction of scientific research activity. Secondly, on the commercial level, research is increasingly undertaken in locations removed from the inquisitive gaze of potential competitors and, consequently, of fellow scientists. It seems to 'disappear' into laboratories 'without windows'.

Following a review of how scientific activity has been modelled in early sociological literature, the chapter begins with a discussion of the development of strategic science and its location in the 'pre-competitive gap' between science and industry. This, it is suggested, provides an ideal research site for an investigation of the different understandings of science held by those involved in its activities. Four models are developed in the context of three different case-studies, and the ways in which science can be said to be 'disappearing' at the conceptual and commercial levels are explored. Elsewhere in this book the phenomenon of 'disappearing science' has been discussed in terms of its invisibility to citizens within everyday life. Here, we observe the different forms of 'disappearance' where science no longer appears in its conventionally portrayed form. The chapter concludes with the view that scientific activity still continues apace, but needs reinterpretation as conventional views are replaced by more sophisticated understandings of what happens in science today.

Models of science

Studies of the social organisation of science have contributed to our understanding of how the definition and interpretation of what constitutes scientific activity is to a certain extent dependent upon its location and social context (see, *inter alia*, Gouldner, 1957; Merton 1957; Kornhauser 1962; Glaser 1964; Ellis 1969; Cotgrove and Box 1970). Kornhauser, in an early contribution, observed: research may mean prestige to some top managers, a tax dodge to the controller, a customer service to the Sales Department, trouble-shooting to the Manufacturing Departments and new products to the Vice President (Kornhauser 1962). This observation was confirmed by Ellis (1969) who similarly concluded that it was possible to determine a multiplicity of perceptions formed by the individual's socio-economic location. However, this finding contradicted the prevailing sociological orthodoxy which relied on a dichotomous view as illustrated by the work of Gouldner (1957), in his discussion of 'cosmopolitans' and 'locals'. These concepts, which were current for some years, were subsequently incorporated into a further theoretical framework, best exemplified in the work of Cotgrove and Box (1970). Their more sophisticated model classified public, academic, and private scientists as being 'professionals', and scientists in large organisations as being (perhaps rather pejoratively) 'non-scientific', despite being actively engaged in the pursuit of science. Here we see the reflection of a more widely held cultural view of the nature of industrially based scientific activity than that of Kornhauser or Ellis, namely that this hardly constitutes 'science' at all. Only those activities that relate to basic, pure, or blue-sky research are appropriately labelled as 'professional' science, thereby contributing to the disappearance of strategic and applied research into the 'non-scientific' category. Equally significantly, however, these approaches suggested that scientists clearly form a heterogeneous rather than homogeneous group, with possibly differing views about the nature of their activities.

Following studies of controversies in science (Collins 1981), it has become clear that science is often best investigated when its basic assumptions are being questioned or challenged. The *pre-competitive gap* in which strategic research is located provides just such a research site, since its principal justification is neither basic nor applied, but a combination of the two in creative tension. Scientists and scientific administrators working in the area of pre-competitive research recognise that its practical meanings are by no means either agreed or shared in the same way, as Cotgrove and Box suggest occurs in the two groups they identify. Many of our respondents have generated their own 'independent' or contextual interpretations of the term 'strategic'. (Quotations from interview transcripts are numbered to prevent identification.)

> Well I suppose a certain amount of [our work] . . . is strategic in the sense that it's being investigated so that it can be applied in the near future and I would say that most of it would be applied within five or ten years certainly, but I would guess the bulk of the work that we actually do is strictly applied research for applied science.
>
> (006: NSP)

> Well I suppose I think of it as a three year time scale – that seems to be the sort of time which people who fund strategic research fund people for – that would be my best guess.
>
> (018: NSP)

> A lot of oceanography would fall into strategic, it is driven by curiosity, you want to know what is going on there, but it is of strategic importance to the UK, from our point of view anyway. I don't know whether the civil servants and the politicians see it the same way but from our point of view we regard it as strategic rather than fundamental . . . well my definition of strategic is not the same as the government's.
>
> (013: NSP)

> I mean strategic research in a purely Thatcherite way is something that you can sell, so if somebody is prepared to buy it from you then it's strategic but by that definition you see NERC is buying our research, and NERC is buying research in universities which people would call pure, so there is a customer for that pure research which is the Research Council, so that's why I find it very difficult to say just where this boundary is. That's why I said when you first came, I'm not sure what strategic research is, it's very difficult to define.   (017: NSP)

These views suggest that particular, personal knowledge of the science may be more important for individuals than general scientific knowledge applied across the field as a whole. Individual researchers, whatever their role, may relate more closely to the specific nature of the issues they face, and reflect this in their understandings of the research process when questioned from outside. This may help explain why, when discussing their own strategic research areas, researchers reflect a variety of views about what constitutes strategic science, and what constitutes scientific activity on a more abstract level.

Our findings clearly show that the three different settings which we investigated, the Protein Engineering Club, the North Sea Community Project, and the Food and Agriculture Research area, generated quite different views of how strategic research is scientifically constituted. They also show that members construct different models of science reflecting their different relationships with each other and the field of study. These are explained in more detail below, and may represent specific 'mental models' (see Adams, Glasner, and Rothman 1988, and chapter five in this collection) held in conjunction with the more widely accepted understandings of science. They form a 'bricolage' held together for social and contingent reasons having been constructed within particular strategic research areas.

We found, for example, industrial respondents working in strategic areas employing a type of discourse which labelled science as a commodity or an enabling and productive force vital to sustaining their company's profitability. One industrial research director (012: PEC) referred to the scientific community in Britain as an 'underexploited resource' with the government being 'absolutely right to try and force that interaction' between industry and academia. Another spoke of the need to 'keep the academics in line and to steer the research in such a way that the companies benefit from the research' (001: PEC). This view contrasts sharply with the notion of 'unfettered science' which is so often presented by elite scientific bodies such as the Royal Society.

Minding the gap

The view that science should somehow be 'harnessed' and channelled towards meeting society's 'needs' has long been debated. Perhaps the most ardent advocate of such a position was Bernal (1939: 415) who perceived the practice of science as:

> The prototype for all human common action [and, furthermore] . . . The task which scientists have undertaken – the understanding and control of nature and of man himself – is merely the conscious expression of the task of human society. The methods by which this task is attempted, however imperfectly they are realized, are the methods by which humanity is most likely to secure its own future.

Following the report of the Haldane Committee (1918) British governments have accepted the need to provide their decision-makers with access to scientific research, and since the Second World War have increasingly sought to ensure the benefits of R and D to a wider community, especially British industry. Yet, as numerous commentators have affirmed (for a recent discussion see Clutterbuck and Craimer

1988) the results have often proved unsatisfactory. It has been argued that for much of the post-War period British governments have pursued the wrong economic goals, weakening British industry. Further, as Roger Williams (1988: 133) points out, many of the reasons advanced in the literature for the decline of British industry 'have at least an indirect connection, and sometimes a very direct one with science and industry'. Williams cites, for example, arguments stressing:

> the anti-industrial ideas transmitted from elites to the public at large via the public schools and the media ... an exaggerated concentration on science as knowledge to the detriment of science as instrument, coupled with the poor training, rewards and status of the applied scientist and engineer in British society. (p. 133)

In summary says Williams:

> for whatever detailed reasons, though highly successful in basic science, Britain has been insufficiently innovative and entrepreneurial, both in comparison with the United States and more comparably in Europe, and also latterly when faced with challenges from Japan and the newly industrialising countries of South East Asia. (p. 133)

Various attempts by Government have been made over the years to improve the efficiency of British R and D. A notable reform was that based on the recommendations of the Rothschild Report (1972), in particular its recommendation that the 'customer–contractor' principle be applied to all applied R and D funded by Government. That is, 'R and D with a practical application as its objective must be done on a customer contractor basis. The customer says what he wants, the contractor does it (if he can), and the customer pays.' Rothschild argued that;

> however distinguished, intelligent and practical scientists may be, they cannot be so well qualified to decide what the needs of the nation are, and their priorities, as those responsible for ensuring those needs are met. That is why applied R & D must have a customer. (Rothschild 1972).

Applied R and D and basic research were distinguished by their end results. That of applied research might be a product, a process, or a method of operation, whereas that of basic research was an increase in knowledge. Whether or not this might turn out to be useful was a matter of chance or, as Rothschild described it, 'a form of scientific roulette'.

Over the two decades since the Rothschild Report, it has become increasingly clear that scientific roulette can provide huge winnings, but that such luck favours the prepared. For example, basic academic research in molecular biology laid the basis for new biotechnology firms. Britain, however, has failed to benefit industrially in proportion to its contribution to the basic research. Research is in reality a continuum, and national research funding policies based on a sharp distinction between basic and applied research have come to be seen by many analysts (House of Commons 1975, 1976; ABRC 1979; ACARD 1985; ACOST 1990) as exacerbating a traditional British problem, rooted in a lack of consensus between industry, and academic research institutions, where most basic research is conducted. In consequence policy-makers began to draw upon the concept of 'strategic research' (ACARD 1985) to handle this widening 'gap'. Crude operationalisations of definitions of basic and applied research had only served to exacerbate the apartheid between industry and academia.

In practice the concept of strategic research has been used to cover research with perceived potential for exploitability, although not necessarily of immediate utility to customers. Moreover, in terms of its location in the research spectrum it can be quite fundamental in nature, although it has a less open-ended time scale than basic research since it is conceived to achieve strategic economic (or military) objectives. According to Irvine and Martin (1984) these may have emanated from two directions, first 'market pull' – occurring when a potential user sees that more background knowledge in a particular field is needed and, second, 'technology push' – when research workers have recognised that a discovery may culminate in some form of tangible application.

Although the strategic research concept incorporates a more proactive view of science in comparison with the traditional, Polanyite (Polanyi 1962), model of science, there remains none the less, the problem of what happens if there is little or no enthusiasm from industry to take advantage of the strategic areas of scientific development identified by government. Also, what happens when the interests of commerce point in different directions to those of science? The resultant lacuna has been labelled the 'pre-competitive gap' (Inter Research Council Co-ordinating Committee On Biotechnology 1982). Successful research organisations, it is assumed, take up these new strategic areas and eventually become involved in an altogether different activity, 'near market research'. However, the precompetitive gap provides an ideal research site for investigating whether different conceptions of scientific activity exist within strategic science.

Case-studies

In order to explore the ways in which different understandings of science are developed within strategic research areas three case-studies were chosen: the Protein Engineering Club; the North Sea Community Project; and the Food and Agriculture Research area. Each had rather different characteristics in terms of formal structure and purpose, but involved scientific 'workers'. The Protein Engineering Club was a joint initiative between industry and academia, with the primary objective being to facilitate the identification of strategic research areas earlier than might otherwise be the case, within a broadly single disciplinary area. The North Sea Community Project was a multi-disciplinary, collaborative research area, funded by NERC with a relatively small input from industry. The Food and Agriculture research area did not involve a single programme, was funded primarily by AFRC except for a small proportion of near-market work, and was limited to a narrow range of disciplines. All three, therefore, were established to identify and pursue strategic research in different ways. This provided a range of contrasting settings within which debates about the nature of strategic research could illustrate different understandings of the scientific process.

The Protein Engineering Club (PEC) was launched in April 1985 as a result of the SERC Biotechnology Directorate's decision to identify and promote an area of science with major strategic potential. It was designed as a research programme for pre-competitive research in protein engineering, and involved new management methods for establishing better co-operation between academic scientists, working within the fields of molecular biology and X-ray crystallography, and their industrial counterparts.

The Club was funded primarily through a combination of the SERC Biotechnology Directorate and four major industrial organisations: Celltech Ltd., Glaxo Group Research Ltd., ICI plc, and J. and E. Sturge Ltd. Each of the four industrial partners, despite committing relatively small sums to support the club's operation, had two representatives on the club's Steering Committee. This arrangement enabled them to wield substantial influence over the club's operation. SERC's total financial input for the first phase was £600 000, while each of the industrial members contributed £30 000.

Our second area of investigation involved another kind of strategic science initiative, the North Sea Community Project (NSP) which was to run for five years from 1988. It was one of five such Community Research Projects run by NERC. It was designed to facilitate collaboration between marine scientists working in universities, polytechnics, and NERC laboratories on the construction of a numerical model of marine quality for the North Sea. As a Community Research Project, NERC co-ordinated funding and facilities, which in this case included contributions from the Ministry of Agriculture, Fisheries and Food (MAFF), and the industrial sector. The North Sea Programme represented the NERC commitment to the support of basic and strategic research in the environmental sciences, although, as the programme developed, the research was focused on assessing the impact of pollution on the North Sea. NERC also failed to generate much funding from private, commercial sources.

The AFRC study, unlike the first two case-studies, did not focus upon a single club or programme of strategic research. Instead a series of interviews were held in various AFRC institutes and research stations involved in such areas of research as: food and crop production methods, public health (for example, diet and nutrition), food quality and preservation, and food bio-technology. Each location visited was funded by a grant for AFRC to a total of 80% of its budget. The remaining 20% was to be secured from external sources, mainly to pay for research classified by AFRC as 'near market'. During the 1980s, AFRC initiated a series of major restructuring changes. The overt political intentions included the promotion of greater co-operation and collaboration between AFRC laboratories and private industry through identifying areas of strategic research. The actual results were, however, controversial, since it has been suggested that the 'public-good' related research – seen as 'near market' research – has been increasingly neglected in an attempt to identify more strategic goals, in the context of a strengthening relationship between industrial firms and the AFRC's research institutes.

Modelling science

For the purpose of our case-studies, we interviewed academic, research council, and industrial scientists, industrial R and D managers, and research council and governments officials. Inevitably, there were some overlaps and ambiguities. Nevertheless, we identified four distinctive models of science in operation. These can be labelled the *Republicans* with views of science very similar to Polanyi's position in his article 'The republic of science'; the *Marketeers*, who adopt a market-orientated view of science; the *Pre-competitors*, those scientists who readily respond to, and perceive as unproblematic the notion of closer collaborative ties with the private sector; and the *Facilitators*, the administrators of science who work either within the research councils or government departments. However, it was by no means the case that these models are held by individuals to the exclusion of any of the others. Interestingly enough, it was possible for an individual simultaneously to aspire to more than one of these, although, usually, one that reflected the structural setting in which the respondent was dominant (Glasner, Jervis, and Rothman 1989).

The Republicans

The traditional view of science was lucidly expressed by Michael Polanyi (1962). Grove (1989: 1) summarises it thus: 'The primary motive for doing science is curiosity and a passion to know . . . [and] . . . to increase our understanding of the world: what it is made of, how it is put together and how it works.' Scientists are idealised as a body of explorers striving for enlightenment, and the intellectual values of science are cross-checked and maintained by scientists as members of overlapping groups. Scientific knowledge is consensual rather than individual, and scientific merit is judged by peers. Intellectual freedom guarantees that new knowledge is discovered; that scientific advance is unpredictable, and that its practical benefits are doubly uncertain. In other words the benefits are contingent upon an unpredictable advance.

According to Baker (1978: 385), Polanyi defined the essential features of freedom in science as: 'the right [of the scientist] to choose one's own problem for investigation, to conduct research free from any outside control, and to teach one's subjects in the light of one's own opinions'. Elsewhere Baker suggests that Polanyi argued:

> with all our might and at every point . . . We must reassert that the essence of science is the love of knowledge and that the utility of knowledge does not concern us primarily. We scientists are pledged to a higher obligation to values more precious than material welfare.

These views have been advocated by numerous commentators on science, especially in the tradition which sees science as a social system (as discussed earlier), reflecting their own practice and a tradition stretching across many generations. It is a view that is currently held, quite widely, in a loose form by many of those who participated in our study, most especially the laboratory scientists. However it is only one of the views of science which were advanced within interviews.

Those who did hold this view among the interviewees suggesting that science is still seen as the 'path to enlightenment', a sort of 'mad pursuit' (Crick 1990), were in a minority compared to the others when it came to deciding upon which aspects of strategic science were to be developed. This Republican view is perhaps best illustrated by the following interview extracts:

> The argument is always that science has a long term investment to the nation so even if it's very blue sky, highly fundamental, it could conceivably be very useful.
>
> (004: NSP)

> I have no time for that [i.e. strategic science] . . . because the contractual research approach means that you know what the answer is pretty much. (016: PEC)

The Marketeers

Predictably perhaps, the Marketeers were committed to the pursuit of commercially exploitable science. Of the utmost priority was the need for science to be evaluated in terms of its economic potential for enhancing the company's profitability within a relatively short time scale. One senior R and D director favoured the adoption of 'a much harder line' towards the academic scientists and close monitoring of the sums contributed by industry. His views of the value of the club concept, as expressed in PEC, had been somewhat mixed. Overall he felt the PEC had failed to produce research of any important commercial consequence. Indeed he believed a comparable sum spent by his company in-house or on a one-to-one basis might well have yielded better results. In his view the research councils still needed to produce more compelling evidence to demonstrate the effectiveness of the club concept.

Furthermore, Marketeers argued that collaborative research should be essentially undertaken to underpin or enable particular types of technology, otherwise it could serve no useful purpose to industry. In particular, industry needs to 'get an early look at something which may give [intellectual] property and then give products – so it has to have something clear in it' (012: PEC). There was an appreciation of the need to embark upon strategic research embodying fundamental and applied research, provided it matched the strategic objectives of the company, and was likely to culminate in a tangible outcome.

The Pre-competitors

Members of this emerging group of individuals were identified within each of the three case-studies. We perceived Pre-competitors to be academic scientists who, having acknowledged the importance of bringing external requirements to bear upon science, sought to undertake 'applicable' research, and were prepared, because of its commercially sensitive nature, that it be subject to some form of constraint. In addition they perceived the inclusion of non-scientists in the process of determining research priorities as both necessary and unproblematic, given that they were materially contributing to the research.

Furthermore, Pre-competitors believed academic scientists ought to understand better the needs of industry and sought to establish a more co-operative set of relationships with their commercial counterparts. 'So as long as we understand each other and what our different needs and aims are then I don't see why we can't get on extremely well' (003: PEC). The Pre-competitors were critical of their Republican colleagues who purportedly have stubbornly remained inside their 'ivory towers' and failed to go out to meet their industrial counterparts. Similarly, they did not accept the Republican's assertion that collaborative research has weakened academic autonomy: 'I think there is an awful lot of hype . . . about that' (003: PEC). Significantly, they were often critical of the opinion, common among the Republicans, that British science is currently under-resourced. One scientist commented: 'I think if scientists have the right attitude and go out and talk to industry then industry listens and gives us money. I'm almost in the embarrassing position at the moment of getting too much support if I wanted it' (011: PEC). Another respondent insisted that academics do need to think more commercially: 'there's every reason why we should learn to exploit the good things and have some fun doing the fundamental stuff – I don't see why you can't do both – certainly that's my aim – to do both' (023: PEC). The Pre-competitors also welcomed the drive by government to make science more responsible to the market requirements of industry, with some referring to 'an enormous amount of waste going on in the academic community. [and denounced the peer review system as] A small club of people who get well financed by . . . wheeling and dealing' (023: PEC).

Finally there was a broad consensus among this new breed of entrepreneur scientists that science actively thrives in a competitive climate because it has helped 'sharpen people up' and has encouraged scientists to publish their research sooner for fear of losing ground, although this may not always be possible as we show below.

The Facilitators

The Facilitators' view of science is as externally directed, for strategic and political reasons, with the overall goal of developing enabling technologies, in order to facilitate take-up by industry. Dissemination is limited by political and economic constraints, and judged in terms of exploitability. Central funding, either in academic or industrial or government department settings, is administered rather than managed, and outcomes are judged by the degree of industrial take-up, within specified time spans.

Respondents advancing such beliefs were predominantly found either within the research councils or government departments such as the Department of Trade and Industry. Significantly, few were prepared to offer their personal opinion, choosing instead to provide an account of their organisation's official policy.

While emphasising their distance from the scientist's bench our respondents continually stressed the need for science to contribute to the advancement of national wealth. Their own responsibility was to ensure that such objectives were achieved through government policy.

One official likened the role of the research councils to that of a scaffolder who erects a temporary structure upon which industry and academia can build. At all times, he continued, it is important for the research councils to 'consider the external environment which includes funding and . . . restricted manpower, [adding that given] . . . there is only a finite number of scientists [and] . . . an even smaller number of world class scientists, restricted resources [and] . . . restricted facilities then inevitably . . . you are into priority setting' (005: OFF). Another senior ex-official described how it was vital for the facilitators to continually 'look forward and anticipate' national requirements, while at the same time promoting linkages between the public and private sectors of science – hence the demand for strategic research initiatives, which another respondent defined as the pursuit of 'new knowledge or [the] re-working of existing knowledge for the potential benefit to the welfare of mankind' (002: OFF).

The Facilitators, on the whole, believed that British science had undergone a period of cultural re-evaluation during the previous two decades. However, this was not necessarily evil, for, whether scientists like it or not, the state has the ultimate authority and responsibility for all expenditure of public funds. It has to say 'Hey, as this is public money it must be properly accounted for and the best way to account for it is to say you must make what you do with taxpayers' money relevant to the taxpayer . . . I regard that as a perfectly acceptable statement' (001: OFF). Thus the government has rightly taken a more vigorous attitude towards evaluating the efficiency and effectiveness of science.

To achieve this last objective they argue scientific programmes need to be better managed and to this end the Facilitators wished to counteract inertia by encouraging greater external participation by industry, or, as one official put it:

> The problem is that people tend to get stuck in those structures, if there was much more diffusion of people through those structures it would be helpful and these labels [i.e. those defining different types of research] wouldn't matter so much. But it is true, and I don't know how one goes about solving it, that people tend to be industrial scientists or academic scientists and have different languages, different ethos and different defence mechanisms, and that to me is where a lot of the problems lie.
>
> (003: OFF)

To summarise, the Facilitators were largely critical of the Republican viewpoint and sought direct intervention where necessary to bring about some form of cultural transformation within the scientific community. Such a proposition would appear complementary to the views advanced by both the Marketeers and the Pre-competitors.

The case of disappearing science: the conceptual level

The four different understandings of science exemplified in our interviews are not exhaustive. The case-studies have generated a number of particular accounts which reflect the scientists' social and economic location within the strategic research area. However, the models strongly suggest that the conventional, Republican view of scientific activity may only be a relatively small aspect of the ways in which contemporary science functions in the late twentieth century. Our findings indicate that, instead of regarding the conventional view as 'professional science' (as described by Cotgrove and Box 1970), and other forms as *non-scientific activity*, the reverse is likely to be the case. Indeed, a 'professional' scientist is now much more likely to hold to a dominant understanding of science as market-oriented, pre-competitive, or administrative. The communalist view held by the Republican scientists no longer appears to have greatest significance.

An analysis of the career developments of our respondents also confirmed the fact that most of those currently involved in these new professional areas, had at some time in their careers (often during early socialisation) been scientific Republicans. Individuals were able to hold a number of different models in their mind simultaneously by relating them to their own career experiences. However, as our interviews show, their current commitments to scientific activity tended to reflect their social locations rather more directly. Scientists were able to manipulate different aspects of their activities *qua* scientists to develop the best 'fit' with their circumstances.

The implications of these processes, the importance of career trajectories, and the ability to manipulate a variety of conceptual frameworks, are far-reaching in the context of the public understanding of science. The conventional view is unidimensional and homogeneous, and not far removed from the wild-haired, white-coated, and bespectacled gentleman scientist creating new knowledge at his laboratory bench. The reality is very different to this idealised view both in appearance, but more importantly at the conceptual level. What exists is not unity but diversity, not homogeneity but heterogeneity.

It is therefore very likely that for the public at large (and perhaps even for some scientists themselves) science may be 'disappearing' partly because what is recognised as scientific activity is moving from ideal type to stereotype. The strategic research areas covered by our case-studies provide strong evidence that this appears to be the case.

Commercialisation and science

Science is not just disappearing from view at the conceptual level discussed above, but is in the process of shifting from sight in a more obvious way in industrial laboratories and some commercially sponsored university research groups. Before exploring this in detail, it is useful to discuss the policy and other changes which have provided the necessary framework.

Several commentators (Ince 1986; Dickson 1984; Aronowitz 1988; Habermas 1970; Nikolayev, 1975) have in the recent past identified a marked shift in science policy as practised by many Western governments and businesses towards the containment of research both in scope and in funding. Dorothy Nelkin (1984) describes how the problem surrounding patent rights in the US has led to a 'growing number of legal and administrative disputes' which have helped raise 'fundamental issues of professional sovereignty; scientific secrecy and proprietary rights'. She further questions how far governments and industry should be empowered to control and/or restrict research and asks whether the 'necessary safeguards' to protect the scientist's academic autonomy are currently in existence. The crucial issue for Nelkin (which she believes has deep implications for the scientific community) is ownership and control as exercised by the state and the commercial sectors. Because scientific research is increasingly closely linked to the interests of power and profit many 'scientists often find themselves embroiled in struggles over the control of the process of investigation, the data produced and the ideas derived' (Nelkin 1984: 4).

Research undertaken during the 1960s found detailed evidence showing that industry viewed the 'economics of research as secret' (Nikolayev, 1975). More recent evidence, as noted earlier, shows that new organisational and state strategies are being developed in order to constrain the flow of scientific knowledge. For example, the EC Committee on Economic and Monetary Affairs and Industrial Policy in a report entitled *The Economic Consequences of the New Technologies* advocated the need for 'the control of knowledge, which covers patent protection, standards policy and the development of the information and language industries as well as questions of technology transfers and technological autonomy' (Besse 1987: 5).

In Britain, however, there has been a partial recognition of the potential pitfalls of such policy developments. One submission made to the House of Lords Select Committee on Science and Technology (1988) argued: 'Commercial confidentiality must not be allowed to impose secrecy on publicly supported programmes, nor must the totality of such programmes become too influenced by industry's necessary commitment to the market place.' Whilst another contributor to the Committee believed: 'the strategic research concept may well result in an imbalance between basic and applied science, encourage the greater duplication of research and necessitate tighter controls on academic freedoms'.

Despite such reservations, it is clear that the identification of exploitable science has become an important feature of British science policy. Consequently, a series of policy changes supposedly geared to sustaining the nation's economic and industrial competitiveness were introduced. This resulted in a dramatic restructuring of British science designed to encourage, or force, scientists wishing to secure funding to look increasingly to the private sector and/or to collaborative ventures. This shift in emphasis has, according to Irvine and Martin (1984), raised the question of 'who controls what and for what purposes', and has led to certain areas of basic research being redefined as strategic. The possibilities for conflict creation in that trend can be seen from the recent NIH furore in The Human Genome Project over the question of patenting of base sequences (Anderson and Aldhous 1992).

Individual respondents discussed the commercialisation process within the context of their own prevailing model of science. In some cases there were very different views propounded about the significance and importance attached to the issues. None the less it was accepted that science is, in a very real sense, now widely regarded as a commodity. To illustrate this particular point it is worth quoting one respondent at length:

> Ten years ago here in Britain in oceanography, and NERC in general I think, we were [a] much more, academically pure, research organisation . . . one would go to meetings in Europe and America or whatever and people would give papers and you would stand up say, 'Hang on, we've solved that problem and this is in fact what you do, or if you write to us I'll send you the coding, and so on . . .' [We] had a much freer perspective, we were exchanging our know-how very freely with people . . . I think I eventually

realised that what was happening was that it was just that we were from different backgrounds, often the Americans or the Europeans or whatever had similar know-how but they were in a much more commercial environment and were much less free to give it away, in fact they thought we were pretty stupid I think in giving it away. Now we are moving very much into that same area so with competition it does affect you, people who used to write to me maybe a few years ago, organisations in the UK where up until then I saw it as our role very much to help these people as much as you reasonably could, and so if they wanted something you just whipped it out of the drawer and xeroxed it and off it went. Nowadays those same people come on the phone and you tend to think two things, one – how much can I charge them for this, and the other is – why give this away, this same organisation, we could be competing for a contract with them, so why give them a head start on us, so I think that question you asked about the commercialisation of science in the UK and its effect on freedom, it's not on publications in our case, it hasn't got that far at present but it is in terms of interaction with other people and it is regrettable in some ways.   (021: NSP)

Others expressed concern that such a shift in thinking is helping to constrain scientific advancement:

I think any limits on communication must restrict the growth of science because very much in our science anyway where the sort of basics have been known for years and each one of us is just sort of building tiny little blocks on top of other blocks and the individual's contribution is minimal really in total, it's the integration of everything, and when you start dividing the structure up then I think there is a long lead time on that but ultimately yes.   (018: NSP)

Consequently there has been the continuing demise of small science in favour of big science whereby 'big clubs form' with one of the most disturbing features being that individual scientists will become:

less than truthful frankly, you know a lot of people who get up and say they can do things or have done things and all of their friends stay quiet and their enemies attack them and their friends then support them and then in the bar afterwards they admit that what they said wasn't true or whatever and so science becomes a commodity and so just like any industry it has competition and the competition isn't at the same healthy level as I think beforehand.   (021: NSP)

Such a process will cause scientists to 'become fossilised' and 'toe the party line' and help to stifle scientific inquiry:

you know the old business of steam engines will they be more efficient than internal combustion but Ford's have got fifty years of know-how and really don't want to know . . . very much the same thing can happen in science. [Thus such structures] . . . if they become too rigid are harmful to the free exchange of ideas and are not as receptive to new things.   (021: NSP)

Within the context of the PEC a number of conflicting views and objectives were promoted which likewise strengthen the case for seeing conventionally defined science 'disappear'. The principal concern for the industrial members was to ensure that few details of the research (during its initial phases) be made known to other scientists or people in rival commercial organisations outside the club. In particular they were keen to identify 'potentially exploitable' research outcomes and, wherever necessary, to suppress their publication. More than one academic complained that the industrial partners endeavoured to restrict the flow of information between the scientists inside the club itself – an action which 'caused immense friction' during the initial stages: 'people were saying they'd never heard about this, why weren't we told, who was asked, why did you write to him, why didn't you write to me, my Head of Department never circulates letters of this kind, etc., etc.' (007: PEC). One suggestion was that the commercial partners' main interest lay in the development of protein crystallography – largely true according to one industrial participant. This had the effect of splitting that community into two and the creation of 'a sort of inner club' where knowledge was withheld from other club members (015: PEC).

Other scientists were concerned because, at the onset of the club's conception, 'the decisions very often had to be made by people who were not scientists' which caused some academic scientists to believe the industrial partners had secured 'a fairly easy ride' having 'put in very little money' whilst being able to exert 'a major influence on the projects that (got) supported' (109: PEC).

It was suggested that the North Sea Project's primary objective had been to assess the impact of pollution in the North Sea and not to map its currents, a view dismissed by some of our academic respondents who insisted such comments 'smacked of political opportunism'. However, one scientist observed, 'the reasons why the NSP was set up were probably not scientific ones they were political ones' (020: NSP). Another respondent believed the Government's overall commitment to the marine sciences as such to be questionable, especially given the scale of redundancies:

> Well clearly in the NERC area a lot of people have essentially been made redundant – some very good people have gone abroad and it is the same around the whole scientific community I think . . . people are not in very good spirits and a certain buoyancy has gone – particularly in the NERC institutes. (018: NSP)

A number of interesting and diverse arguments were also offered as to why the NERC had largely failed in its attempts to secure for the Project any significant interest or funding from commercial sources. One suggestion was that the Research Council had not approached industry early enough and:

> If they [industry] were going to be involved [then they] should have been involved right at the beginning . . . [Furthermore] What's strategic to NERC may well not be what's strategic to an oil company, and in fact it may well be that there's a time lag there, and the strategic research the oil companies would have wanted would have been years ago. (002: NSP)

Whilst another oceanographer observed the meaning of the term strategic is 'very different, very different for the various groupings involved' (005: NSP). Another difficulty facing NERC was that of designing a multidisciplinary and collaboratively based research programme; largely because 'you can't actually please the two masters' at the same time. Furthermore, the only feasible way to help resolve such problems is to provide for 'a lot of discussion, a lot of discourse, between the individuals and the groups concerned' and not to 'go cap in hand to industry at the end of it all and say you're a bit short. I don't think industry would be that sympathetic, is my general impression' (005: NSP).

It is clear, therefore, that the commercialisation of scientific research has focused the concerns of practitioners at all levels on the potential difficulties it poses for the untrammelled pursuit of knowledge and truth. One particular area of concern is the need for secrecy and the fear of industrial espionage, which is explored more fully in the next section.

The case of disappearing science: the commercial level

A major area of concern among a number of respondents, particularly in the PEC, was academic laboratory–industry co-operation in developing new products or new materials and its impact on the Mertonian norm of disciplinary communism. For the scientists who hold the Republican view of science, co-operation is proper, above board, open for discussion, and, relatively speaking, unproblematic. These views were antithetical to the Pre-competitive scientists who were committed to the pursuit of strategic science and the consequent need for commercial secrecy.

Some of the Republican scientists suggested their work was being 'poached' by commercially minded researchers, while others believed there is a growing tendency for research to be conducted 'behind closed doors' or, as one put it, 'in a lab without windows'. The same senior scientist went further and described how, in his experience, the 'real' research undertaken by one large industrial organisation had been located down a flight of stairs where 'it didn't see the light of day' and was 'certainly not communicated to the other industrial organisations' engaged in the same venture. It was not suggested that this phenomenon had in any sense inhibited the pursuit of strategic science and its later application(s), nevertheless, this did represent an example of the removal of scientific research from the public domain.

The Marketeering view on collaborative initiatives was to ensure that they could participate 'without giving away commercial advantage'. But this was precisely why, in the case of the PEC, a number of the companies chose not to join: 'they didn't want to disclose what they might be working on [and] . . . they didn't want to steer the committee so that other companies could identify their aims' (001: IND). Interestingly one industrialist claimed that certain academic scientists had adopted a much more aggressive position than their business partners in their efforts to commercialise the product. When asked why this might be so he said it was largely due to expediency on the part of the academics because it matched the 'current face of university politics – if you can exploit your research then you're the good guys' (001: IND). However he did add that industry remains primarily motivated by 'the aims of profit and commercial exploitation'.

The Facilitator view was less ambiguous. Indeed one Department of Trade and Industry official put it bluntly, arguing that those academic scientists wishing to participate in one of the collaborative programmes should comply with the various conditions stipulated. He continued:

> I don't think it is unreasonable to expect them to go into the programme well aware of what the constraints on them will be, if they enter the programme they must do it with their eyes open. If we are now talking about transferring technology to industry which is going to try and make profits and create jobs out of that technology it's not unreasonable to expect that information to be kept confidential . . . so I take a pretty tough line with [them] . . . I think. [T]here is no way you can easily reconcile academic freedom and industrial confidentiality. They are incompatible.   (001: OFF)

For the Pre-competitor, the Marketeer, and the Facilitator alike 'no publication equals easier exploitation'. Some of those scientists who had direct experience of working in, or on behalf of, industrial organisations described how they were occasionally dissuaded from publishing the results of their work in reputable journals before the necessary patents had been filed. There were also occasions when the employers found it necessary to suppress publication in the interests of commercial expediency, thereby making it difficult for the scientists to register new knowledge in the traditionally approved manner.

Whilst the need to suppress or delay publication clearly conflicted with the Republican position, there were nevertheless additional differences of opinion. Our research confirmed the existence of conflicting priorities for the role of science among our respondents. Industrialists surveyed in our research were most concerned that any collaborative work should be premised upon the creation of 'money-generating results' (002: IND). Thus strategic research constitutes work 'which is supposed to have an economic basis but is not yet of any direct economic interest to individual companies' (002: IND). However, problems emerged because the members failed to agree on a shared programme of research. Each had their own strategic objectives and were loathe to divulge information likely to reward their potential competitors. The implication of this last point is clear – no company wishes to participate in a research programme where there lies the risk of losing its competitive edge. According to some respondents, this resulted in individual members acting warily and defensively at the joint meetings of the PEC.

Some respondents thought that academic scientists ought to accept that if they choose to participate in collaborative ventures with industry then, essentially, they have sold themselves completely outside the academic community and must thereby comply with the necessary restrictions. This problem, however, remains largely unresolved:

> The two sides treat each other with a fair degree of suspicion and I think in my view, the suspicion is more justified on the industrial side than the academic side. Because they [academics] want to be able to do what they want and publish it all, but at the same time collect money from industry . . . [whereas] the objectives . . . [of] industry in putting in that money is to promote confidential research which they want to keep secret and exploit commercially; and the revelation of those results outside upsets the applecart in terms of exploitation. (002: IND)

Indeed the same interviewee went further and suggested that some academic scientists wished to develop their own commercial interests by putting themselves in 'a sort of middleman position so that they're actually in a position to exploit'. This view was supported by some Republican scientists who believed such an outcome was precisely that anticipated by the Pre-competitors.

Rediscovering science

In theory, the public understanding of science appears to be a zero sum game with a limited number of concepts and a connecting language, which can be grasped by anyone with sufficient wit and enthusiasm. In practice, this project has argued that the range of different models of science available is a function as much of the context and historical location as it is of conceptual structure. In showing this, the research has highlighted the importance of seeing what is missing from discussions of strategic science, and using the gaps, not to discover an abstract notion of traditional scientific discovery, but to rediscover the real world of science which has all but disappeared in discussions of what constitutes an appropriate public understanding of the phenomenon.

It is therefore apparent that understandings of science may suffer because science does not fit with the traditional model as described by Polanyi's 'Republic of Science', not only for those working in strategic areas at all levels, but also for the public at large. It is this traditional model which appears to have 'disappeared' from the discourse of science policy-making to be replaced by a much more heterogeneous collection of differing models of appropriate scientific activity. Science has therefore to be rediscovered in these different models without the preconceptions normally underpinning sociological research.

Our findings show that science, more especially science in strategic areas, appears to be disappearing from various levels of discourse in the public domain, and to be doing so in a variety of ways. However, it also recognises that this may be the result of a magician-like sleight of hand which is only possible because the eyes of the audience are glued to a traditional and relatively narrow model of scientific activity. Science waits to be rediscovered by research which refuses to accept this conventional view.

REFERENCES

ABRC (Advisory Board for the Research Councils), 1979, *Third Report 1976–1978* (London: HMSO).

ABRC (Advisory Board for the Research Councils), 1985, *Science and Public Expenditure* (London: HMSO).

ACARD (Advisory Council for Applied Research and Development), *Exploitable Areas of Science* (London: HMSO).

ACOST (Advisory Council on Science and Technology), 1990, *Developments in Biotechnology* (Cabinet Office, HMSO).

Adams, C., Glasner, P., and Rothman, H. 1988, 'Engineering proteins: public understanding of science and identification of strategic research areas', in *The Study of Science and Technology in the 1990s*, Joint Conference 4S/EASST, Amsterdam.

AFRC (Agriculture and Food Research Council), 1987, *Corporate Plan 1987–1992* (London: HMSO).

Anderson, C., and Aldhous, P., 1992, 'Genome project faces commercialization test', *Nature* 355 483–4.

Aronowitz, S., 1988, *Science as Power* (Basingstoke: Macmillan).

Baker, J. R., 1978, 'Michael Polanyi's contributions to the cause of freedom in science', *Minerva* 16, 3.

Bernal, J. D., 1939, *The Social Function of Science* (London: Routledge).

Besse, J., 1987, *The Economic Consequences of the New Technologies* (EC Committee on Economic and Monetary Affairs and Industrial Policy), Draft Report, WG (VS1) 4060E (Brussels).

Clutterbuck, D., and Craimer, S., 1988, *The Decline and Rise of British Industry* (London: Mercury Books).

Collins, H. M. (ed.), 1981, *Knowledge and Controversy*, special issue of *Social Studies of Science*, 11, 1.

Connerade, J. P., 1988, 'Public perceptions of science: pursuit of knowledge or engine of profit', *Papers in Science Technology and Public Policy* No 18 (London).

Cotgrove, S., and Box, S., 1970, *Science, Industry and Society* (London: George Allen & Unwin Ltd.).

Crick, F., 1990, *What Mad Pursuit: a personal view of scientific discovery* (Harmondsworth: Penguin).

Dickson, D., 1984, *The New Politics of Science* (New York: Pantheon Books).

Ellis, N. D., 1969, 'The occupation of science', *Technology and Society* 5, 33–41.

Glaser, B., 1964, *Organisational Scientists: their professional careers* (Indianapolis: Bobbs-Merrill).

Glasner, P., Jervis, P., and Rothman, H., 1989, *Public Understanding of Science and Identification of Strategic Research Areas*, Final Report to ESRC (Swindon).

Gouldner, A. W., 1957, 'Cosmopolitans and locals', *Administrative Science Quarterly* (December) 281–92.

Grove, J. W., 1989, *In Defence of Science: science, technology and politics in modern society* (University of Toronto Press).

Habermas, J., 1970, *Towards a Rational Society* (Heinemann Press).

Haldane Committee, 1918, *Report on the Machinery of Government* Cmnd 90230 (London: HMSO).

House of Commons Select Committee on Science and Technology, 1975, *Scientific Research in British Universities*, session 1974–5 (London: HMSO, July).

House of Commons Select Committee on Science and Technology, 1976, *University–Industry Relations*, session 1975–6 (London: HMSO).

House of Commons Select Committee on Science and Technology, 1988, session 1987–8, 4th report (London: HMSO).

Ince, M., 1986, *The Politics of British Science* (Brighton: Wheatsheaf Books).

Inter Research Council Co-ordinating Committee On Biotechnology, 1982 (London).

Irvine, J., and Martin, B., 1984, *Foresight in science – Picking the Winners* (London: Frances Pinter).

Kornhauser, W., 1962, *Scientists in Industry: conflict and accommodation* (Berkeley: California University Press).

Merton, R., 1957, *Social Theory and Social Structure* (Glencoe, Ill.: Free Press).

Nelkin, D., 1984, *Science as Intellectual Property: who controls research?* (New York: Macmillan).

Nikolayev, A., 1975, *R & D in Social Reproduction* (Moscow: Progress Publishers).

Polanyi, M., 1962, 'The republic of science', *Minerva* 1, 68.

Rothschild Report, 1972, *The Organisation and Management of Government R & D* (London: HMSO Cmnd 4814).

Williams, R., 1988, 'UK science and technology policy, controversy and advice', *The Political Quarterly 59*, 132–44.

Conclusions

ALAN IRWIN AND BRIAN WYNNE

The fieldwork presented in this volume provides mainly qualitative insights into the ways in which public groups attempt to fashion locally useful knowledges from 'external' and 'indigenous' sources. Most of the chapters analyse the interactions between identifiable social groups and scientific, technical or medical experts. These groups are defined by different parameters; geopolitical-cultural location (The Isle of Man); shared livelihood, culture, and physical place, but with more cultural permeability (Cumbrian sheep-farmers); physical location but a less-distinct culture (residents around major industrial hazard sites); shared experience of medical treatment – either chronic or episodic (hypercholesterolaemia and antenatal patients). Two further chapters analyse responses of more diffuse collectivities to scientific knowledges as experienced in museum exhibitions (Chapter Seven), and in relation to radiation hazards in the home (Chapter Five). Finally, and consistent with the orientation of the fieldwork chapters, two chapters examine new contexts of the contemporary negotiation of scientific practice (i.e. the norms and ethos of what is meant by 'science') – namely in environmental debate and policy-making, and in the commercialisation of scientific research.

In these conclusions, we offer some thoughts on the overall implications of this work by giving further reflection and clarification to some of the key themes of our book. Since two of the key assumptions about science which frame the 'public understanding' issue are its intrinsic usefulness and its universality, it is especially important to give attention to two issues at this stage: the connections between 'useful knowledge' and hidden models of social agency; and the relationships between the 'local' and the 'cosmopolitan' in the 'micro-social' research presented here. Following this, we will consider some of the practical lessons which can be drawn from our collective research.

Useful knowledge, social agency, and legitimation

'Useful' knowledge in this context means valid and socially legitimate, as well as being of more immediate practical relevance and use. It is often found that expert knowledge has been ignored by social groups because it is not tailored to the needs, constraints, and opportunity structures of the social situation into which it has been interjected as authoritative knowledge. Emergency planning information which is insensitive to the practical differences between daytime – when family may be at school and work – and night-time – when they are likely to be together at home – is one such example. Our collection has suggested many more.

This book has already implied the need for sensitivity to the 'local' context(s), and for listening to and understanding 'user' situations and knowledges. However, the legitimacy of external expertise also involves further dimensions. These include its accessibility and accountability, whether it implies empowerment or disempowerment of social actors, and its consistency or otherwise with relevant cultural idioms. As the case-studies clearly show, these dimensions amount to more than questions of 'presentation' or 'communication' of what are assumed to be already-validated knowledges.

The social legitimation of expertise necessitates the reopening of expert knowledge and its validation all over again – but in more complex, less reductionist circumstances. Often, as the case-studies have shown, the prior context of scientific validation has been shaped by social assumptions (for example, about user-capabilities and needs) and these have been 'black-boxed' into analysis. Thus, social assumptions within science become exempt from negotiation. This raises questions directly about the contexts of validation and construction of expert knowledges. Repeatedly, we find expert knowledges (and associated social programmes, be they emergency planning, environmental protection, medical treatments, or public education) running into difficulties because they have assumed that validation of expert knowledge is completed before (and insulated from) its social deployment or use. The shaping social assumptions then become problematic prescriptions – inviting public alienation from scientific discourse.

So deeply ingrained is this assumption that all of the troubling experiences of apathy, resistance, plain distortion, and exaggeration which disfigure the public life of science in modern 'scientific' democracy have led to little or no consideration of whether they imply anything might be wrong with the organisation, control, and conduct of 'science' (in addition to just its 'communication'). It is taken for granted that the contexts of validation of use or dissemination are separate, and that the source of the problems lies only in the latter domain – partly with the inability of scientists to communicate their knowledge, but mainly with the inability of publics to assimilate it.

Thus the 'public understanding of science' problematique has been constructed in such a way as to project onto 'the public' the internal problems and insecurities about legitimation, public identification, and the negotiation of science's own identity which pervade scientific institutions. This has the effect of pre-empting complex and threatening questions about contemporary institutional structures and cultural properties of science, and their correspondence or otherwise with broader cultural and social norms. Questions pertaining to legitimacy and the backcloth of public uptake of, or alienation from, science thus become obscured.

Within this dominant ideology, 'the public' has been constructed in particular ways which our research challenges. For example:

A 'the public' is usually implied to be an aggregate of atomised individuals with no social composition, hence no legitimate autonomous cultural substance;

B ignorance – defined as the lack of knowledge as measured against scientific standards – is taken to be a function of intellectual vacuum or incapacity, whereas (Chapter Five) our research shows it often to be a function of active reflection upon, and construction of, the actor's *social* position and identity in relation to scientific-technical institutions;

C it is assumed that the actor's basic values are identical with those of science – for example, that she is concerned to maximise control, rather than perhaps to negotiate and adapt to actors and forces recognised to be beyond such control or which should be beyond such control. Hence, the epistemological commitments which frame science, namely instrumental control, are assumed – wrongly – to be the automatic norm defining all valid knowledge;

D lay people are assumed to desire and expect certainty, and risk-free environments, so that their lack of enthusiasm for science can then be attributed to their alleged inability to face up to science's 'grown-up' recognition that risk and uncertainty are intrinsic to everything. Yet our research reinforces previous work in showing no such naiveté on the part of the public; indeed it shows the central kind of risk being faced as that of dependency upon increasingly expert-imbued social institutions, the basis for trust in which is obscure;

E the public is assumed to exist within the same structure of social agency as the expert or anyone else, in that the social opportunities for (and implications of) use of available knowledge are assumed to be homogenous throughout society. This is a corollary of point A above. In effect, it denies structures of power and dependency in society, and the inscription of these into people's 'natural', culturally rehearsed boundaries of possible action and knowledge.

The fieldwork reported here roundly contradicts each of these articles of faith in the dominant worldview which shapes the conventional public understanding of science agenda. This orthodox construction of 'the public' and 'public understanding' systematically deflects attention away from critical debate about science and scientific institutions, about the ownership and control of science and its products, and about the implicit social visions these carry. It is an important finding from our research that, since public experiences of science can never be detached from imputed institutional interests and agendas of whatever kind, the manifest *lack of reflexivity* on the part of science in public only amplifies any existing tendency for public groups to mistrust it.

It follows from our argument in this book that 'useful' scientific knowledge needs to be reflexive and self-aware rather than dismissive of such social and epistemological concerns as irrelevant and 'soft'. If science is to work with rather than against public groups (or simply be ignored by them), then 'usefulness' and 'self-reflexivity' must form part of the same social and institutional processes. What is meant by science in given cases must be more open to structured reflection and negotiation, with particular attention to the conditions of validity of the relevant knowledges.

The local and the cosmopolitan

The unreflexive character of science in public is encouraged by a further misconception of the relationship between the 'local' and the 'cosmopolitan', and, paralleling this, between the 'micro-social' qualitative research presented in this volume and the 'macro-social' quantitative data of public understanding of science survey research.

A common way of describing the relationship between large-scale quantitative surveys of public understanding of science and micro-social qualitative research is to present the former as providing the objective large-scale picture of the structural realities of public understanding, and the latter as filling in the details. Thus, at the outset of the ESRC/SPSG public understanding of science initiative from which this work is derived, but which also supported some large-scale quantitative public surveys, there existed an assumption amongst some that the role of the qualitative studies was to help elaborate the survey questionnaires by finding richer detail and possible differentiation about likely lay sources of information, trusted expert bodies, and local uses and practices. Questions about the meanings of 'science' to different social groups, and the possibility of unarticulated epistemological conflict and negotiations about the implicit social purposes of knowledge, were simply not conceived to be issues within this perspective. In other words, 'universal' science was not problematised. Qualitative social research on public understanding was seen only as intellectual embroidery within the 'objective' macro-social patterns revealed by large-scale surveys. Put differently, large-scale surveys would offer the overall black-and-white pattern, whilst ethnographic projects would provide occasional splashes of technicolour.

We regard this as a fundamental misreading of the relationships between the (micro) qualitative and the (macro) quantitative, and between the local and the cosmopolitan empirical focus in such research areas. Consistent with the developing body of social-theoretical debate about modernity and post-modernity, we note that the social processes of response to and (re)construction of science can be understood, not by seeing them merely as local embellishments of life within the objective cosmopolitan parameters of scientific modernity, but as responses partly to the fundamental failures of those 'universal' modernist institutions, programmes, and promises even within their own instrumental terms.

Thus, to take an example from Chapter One, Cumbrian sheep-farmers were not predisposed to refuse credibility root-and-branch to scientific advice about the Chernobyl fall-out. However, they soon found that they were being offered no opportunity to negotiate the conditions of validity nor the boundaries of the science in their situation, thereby denigrating their own specialist knowledge and social identity. The science was not fulfilling its instrumental claims. Furthermore, it was culturally alien in its epistemological assumptions of predictive control and determinacy, rather than the adaptive flexibility and indeterminacy which represented a more familiar cultural idiom to the farmers. There was, in other words, a deep and unarticulated conflict about the proper scope and nature of human agency, a social–cultural conflict which was not even recognised. In so far as institutionalised science reflected an assumption of control, this was also a conflict over the political boundaries of scientific expertise in relation to their accustomed structures of decision-making.

In other words, we would argue that the 'local' machinations around science as analysed in this collection are of much wider significance than the particular local context in which they are manifested. This is true for each such case, setting aside for the moment the aggregate significance of the countless similar cases we could have examined. In different detailed ways, they all witness the attempted renegotiation of the social and cultural reach of modern scientific culture, of the epistemological principles of instrumental control which have become enshrined and taken for granted as the 'natural' criteria of valid public knowledge, hence of privileged authority, in modern society. In many of the cases in this volume, lay people operated effectively and constructively, despite obstacles, as peer-reviewers of the validity of scientific expertise applied to their situation. This extended peer-group role extends into negotiation over the legitimate reach of scientific principles of instrumental control, and of the interests seen to be shaping particular scientific claims and interpretations.

Thus, 'local' processes and cases as described in – and only accessible to – qualitative research methods are of far wider significance than conventionally supposed, because they are prototypical renegotiations of the basic nature and legitimacy of cosmopolitanism, that is of modernism and its assumptions of the universal remit of instrumental rationality and good science as we know it. Although in political discourse, such unitary and homogeneous models of good science are deployed as standards of rational authority, empirical examination shows practice to be far more flexible, diverse, and indeterminate. Jasanoff's account of the negotiations of practically effective definitions of 'good science' for environmental risk management amongst the US regulatory agencies is a case in point here.[1]

The redefinitions of science under commercial prerogatives as described in Chapter Nine also exhibit indeterminacy and indeed confusion about ultimate normative models of science, whether or not this relatively private fluidity is ever reflected in more public debate. Likewise Yearley's chapter on environmentalist constructions and uses of science indicates a similar indeterminacy about what is to count as valid knowledge in 'the' public sphere. If the social purposes of public knowledge are to uphold less instrumental and exploitative relationships between human society and nature, and between human beings themselves, as much of the impetus of new feminist, environmental, and other social movements and post-modern critique would claim, what does this imply for the redefinition of 'science' as valid public authority?

Regardless of the concrete answer to this pressing question, we argue that the fundamental social dynamics of many 'ordinary', so-called local cases of public response to, and negotiation with, scientific expertise are similar to those of the more articulated cultural critiques of modernity. The local is the site of renegotiation of the 'universal'.

Much of the interest in public understanding of science stems from the field of risk, often arising from frustration amongst experts that the public is apparently unable to comply with what the experts say about risk magnitudes. The simplistic idea that people should compare quantitative estimates of risks and shape their acceptance of technologies accordingly, has given way to a more developed understanding of the social rationality of expert and public risk definitions. Whereas expert risk assessments usually take the trustworthiness of relevant institutions for granted, this is often precisely the focus (explicit or otherwise) of public concern and scepticism. Indeed broader theories of 'the risk society' have identified a pervasive public mood of scepticism and alienation from the dominant institutions of modern society, most especially those like science which claim unconditional and universal warrant as the epitome of modernity.[2]

What scientists interpret as a naive and impracticable public expectation of a zero-risk environment can thus be seen instead as an expression of zero trust in institutions which claim to be able to manage large-scale risks throughout society. These social dimensions of risk and trust are the general context within which specific issues are played out. Action, organisation, and the informal negotiation of collective loyalties outside the range of formal established institutions is an increasing phenomenon of contemporary society, and to the extent that modern institutions do not recognise that the question of trust is focused on them, they only contribute further to the same 'post-modern' trends.

For as long as universal science is unable to recognise and accommodate such fundamental pluralism (in the context of ownership, control, and validation, as well as just of 'application') as legitimate and necessary, 'public understanding of science' (for which read public identification with science) will remain a dismal story of failure and retreat. In so far as it can begin to recognise and reflect openly upon its own deep cultural biases, science may find the latent heat of evaporation of the public understanding problem to be surprisingly low. The politics of legitimation may be best conducted by questioning the anxious culture of control.

Some practical considerations

As a final section to these Conclusions, we now need to consider directly the practical and policy implications of our analysis. Of course, individual chapters have made detailed suggestions of their own about this. At this stage, we can simply emphasise some of the more general points.

In particular, the analysis in this book has led us away from a problematisation of the public as 'misunderstanding' the true nature of science, and towards the notion that contemporary science and scientific institutions 'misunderstand' their own epistemological limitations but also the public (or 'local') contexts within which they must operate. This will be an uncomfortable conclusion for the scientific community, since it suggests a pressing need for debate over the limitations of science as well as its putative benefits. However, in a situation where public groups more often see science as an *obstacle* to development rather than a facilitator, there may be little choice – unless we are to maintain the present largely sterile discussion between the defenders of science and those who would wish to reject it for its disenchantment (and obfuscation) of the world. In making this argument, it is also important to recognise the changing cultural context within which science must operate – where claims to authority are likely to be met with an increasingly critical (if not downright hostile) audience. As Giddens has expressed this:

> Lay attitudes towards science, technology and other esoteric forms of expertise, in the age of high modernity, tend to express the same mixed attitudes of reverence and reserve, approval and disquiet, enthusiasm and antipathy, which philosophers and social analysts (themselves experts of sorts) express in their writings.[3]

Scientific institutions thus need to go forward with a full recognition of this complex and dynamic social setting, rather than indulging themselves in versions of the 'deficit theory' which will prove *un*-productive and even *counter*-productive.

Going further, a major aspect of this process will involve the recognition not only of the limitations to scientific forms of understanding, but also of the existence of alternative and more 'local' forms of knowledge and knowledge practice. Whilst these are sometimes known as 'contextual knowledges', it is important for us to note that science also represents a form of contextual knowledge – the context in that case generally being one of *ceteris paribus* assumptions and of laboratory-controlled conditions. Once again, rather than attempting to maintain a knowledge hierarchy, the aim should be to acknowledge and build upon this broader network of knowledge relations – always accepting that together they can represent a rich and well-tested body of contextual knowledges.

In all of these steps towards more progressive relationships between knowledge and citizenship, it will be particularly important to consider the emergence and development of new institutional forms which attempt to deal with these issues in a progressive and imaginative fashion (for example, by bringing together both 'indigenous' and 'formalised' knowledges). Whilst it is not necessarily the case that current 'social experiments' such as those involving science shops, local environmental campaigns, feminist networks, trade union activities, and tenants organisations will consciously address the kinds of issue about science and citizenship which this book has raised, it may well be that there are important lessons to be learnt here. Whatever the case, we would argue that such initiatives need to be considered within the kind of analytical and conceptual framework presented in this volume. They might then be constructed as 'learning systems' in the fuller sense implied here.

Quite clearly, such localised and specific initiatives struggle to gain credibility within scientific institutions – being seen as the murky world of politics and direct action, rather than the preferred and more-cloistered world of science (even though the politics of this supposedly cloistered world is just as pervasive). More fundamentally, such social experiments in new science–public interactions often succeed only in re-emphasising the gulf between scientific institutions – and scientific knowledges – and the general public. As one account of a science shop initiative in France puts it: 'I came to feel that I was trying to convince all parties concerned – the public, scientists and institutions – of the credibility of something that none of them wanted. All in all, not a bad version of hell.'[4] As the same commentator notes, achieving a more productive relationship between science and its publics will involve far more than the 'de-mystification' of science – it also, as our research has suggested, requires the establishment and maintenance of progressive relations of knowledge and citizenship. This will, of course, also involve – in proper context – improvements in conventional scientific and technical literacy on the part of public groups.

For all the reasons discussed in this book, it is important that science is prepared to learn from practice in this area rather than seeing it merely as the 'appliance of science'. Equally, it is important for scientific institutions to recognise that science is often seen by public groups as a resource for the powerful in society – and against the everyday interests of the weak. Only deliberate – and deliberately humble – efforts in this area can begin to address the issues. In the current climate, science should play an important role within a social dialogue over socio-technical development – and not simply resort to the arrogance of a supposed 'higher rationality'. Such a worldview only serves to reinforce current attitudes of ambivalence, hostility, and indifference. Even within the framework of technical-economic exploitation of science it is counter-productive.

In final conclusion, therefore, we wish to emphasise the institutional dimensions of science and the significance of these for the social legitimation of science. Thus the practical target of advancing the public understanding of science depends upon a willingness to facilitate a broader discussion of the contemporary – and changing – character of science and the relationship between this and wider relations of knowledge and citizenship. This will raise difficult questions about the limitations of scientific understanding, the direction of scientific research, the relationship between public needs and private profit, and, ultimately, about who should control science. Now that the discussion over the 'public understanding of science' has been initiated – and, at least partly, researched – it is important that it should be released from its orthodox restrictions and developed as a major opportunity for a society-wide debate of a more fundamental kind than has so far been officially recognised (even if it is already taking place in an oblique fashion). Scientific institutions – and individual scientists – have an all-important role to play in these necessary developments, albeit one which differs markedly from that which has dominated so far.

NOTES

1. Jasanoff, S., *The Fifth Branch – science advisers as policymakers* (Cambridge, Mass. and London: Harvard University Press, 1990).

2. See, for example; Beck, U., *Risk Society; towards a new modernity* (London, Newbury Park, New Delhi: Sage, 1992). Giddens, A., *Modernity and Self-Identity; self and society in the late modern age* (Cambridge: Polity, 1991). For a more thoroughly constructivist view of the relationships between expert and non-expert cultures, see B. Wynne, 'May the sheep safely graze? A reflexive view of the expert–lay knowledge divide', in *Risk, Environment and Modernity: towards a new ecology*, edited by S. Lash, B. Szerszynski, and B. Wynne (Newbury Park, Ca, and London: Sage, 1995).

3. Giddens, ibid., p. 7.

4. Stewart, J., 'Science shops in France; a personal view', *Science as Culture* 2 (1988) 62.

Notes on contributors

These notes are organised according to chapter order.

ALAN IRWIN is a sociologist with a special interest in issues of science, risk, and the environment. He is Reader in Sociology at Brunel University. For over a decade he has been involved with issues of risk assessment, science and technology policy, sociology of scientific knowledge, and environmental sociology. He has written *Risk and the Control of Technology* (Manchester University Press, 1985) *Citizen Science* (Routledge, 1995) and various academic publications in this area. Currently, he is working with Steven Yearley on an ESRC-funded project on 'Regulatory Science'.

BRIAN WYNNE is Professor of Science Studies at Lancaster University where he is Director of the Centre for Science Studies and Science Policy (CSSSP) and Research Director of the Centre for Science, Environment and Culture (CSEC). After a first degree and Ph.D in natural sciences, Brian's earliest work in social studies of science (for an M.Phil in sociology of science) was in history of early twentieth-century physics. The distinctive feature of his research approach has been to apply the sociology of knowledge in a symmetrical fashion to both expert and lay knowledges about, for example, risks, and thus to illuminate both as forms of culture which embody situated rationalities. Brian was joint-director of a public understanding of science project at Lancaster. He has published widely on issues of risk, environment, and science.

ALISON DALE was trained initially as a biochemist and now works with issues of science and technology policy. The project discussed in Chapter Two developed out of her postgraduate research into the public dissemination of major hazard information. She is now employed as a Research Associate at the University of Manchester within PREST (Programme of Policy Research in Engineering, Science, and Technology).

DENIS SMITH is Professor of Management at the University of Durham. He is also joint Head of the University's Centre for Risk and Crisis Management. Denis

Smith's main research interests are in strategic management, risk assessment, crisis management, and corporate responsibility. He has previously edited a collection on *Business and the Environment: implications of the new environmentalism* (London; Paul Chapman, 1993) and co-edited *Waste Location: spatial aspects of waste management, hazards and disposal* (London; Routledge, 1992).

HELEN LAMBERT is Lecturer in Medical Anthropology at the London School of Hygiene and Tropical Medicine, University of London. She specialises in the study of health, illness, and medicine in both India and the UK. At the time of the study on which Chapter Three draws, she was a Research Fellow at the University of Bradford working with Hilary Rose on a two-year project. Helen Lambert has published in *Social Science and Medicine* ('The cultural logic of Indian medicine' 34,10 (1992) 1069–76) and (together with Klim McPherson) in a book edited by Basiro Davey and Jennie Popay (*Dilemmas in Health Care*, Open University Press, 1993).

PROFESSOR HILARY ROSE is a sociologist of science who has published extensively in the field of science, technology, and society. She became interested in how patients make sense of the biomedical science relating to familial hypercholestorolaemia after learning that she herself had this genetic disorder. She has published *Love, Power and Knowledge; towards a feminist transformation of the sciences* (Polity Press, 1994).

FRANCES PRICE is a sociologist with a particular interest in expertise and regulation in biomedical innovation. She is a Senior Research Associate in the Faculty of the Social and Political Sciences at the University of Cambridge. Her chapter draws on two phases of fieldwork. The first, funded by the Department of Health, was part of the National Study of Triplets and Higher Order Births. The second involved interviews with men and women attending infertility clinics and their doctors.

MIKE MICHAEL teaches sociology of science at Lancaster University. His chapter draws upon research conducted with Rosemary McKechnie and Brian Wynne on the public understanding of ionising radiation. It reflects his abiding interest in discourse analysis, and is part of a trilogy of papers that have addressed the public understanding of science in relation to the construction of society ('Discourses of danger and dangerous discourses: patrolling the borders of Nature, Society and Science', *Discourse and Society*, 2,1 (1991) 5–28), and of identity ('Lay discourses of science: science-in-general, science-in-particular and self', *Science, Technology and Human Values*, 17,3 (1992) 313–33).

ROSEMARY McKECHNIE is a social anthropologist whose current research interest is in processes of identification in Europe with special reference to scientific and technical issues. She is currently a lecturer in social science at the Bath College of Higher Education.

SHARON MACDONALD is Lecturer in Social Anthropology at Keele University. After carrying out doctoral research on language and identity in the Scottish Hebrides, she became a Research Fellow at CRICT (Centre for Research into Innovation, Culture and Technology) at Brunel University. The ethnographic fieldwork on which Chapter Seven draws was directed by Roger Silverstone, now Professor of Media Studies, University of Sussex. Results of the research have been published in various journals including *Cultural Studies, Public Understanding of Science, Anthropology in Action, Publics and Musées*, and a special edition of *Science as Culture* (for which Macdonald is the guest editor).

STEVEN YEARLEY is Professor of Sociology at the University of York. He has been doing sociological work on environmentalism for the last ten years, though he also does research on the sociology of science. His most recent books deal with environmental topics: *The Green Case: a sociology of environmental issues, arguments and politics* (Routledge, 1992) and *Re-inventing the Globe* (Sage: 1996). He is editor for the Routledge 'Environment and Society' series. Currently, he is studying the internationalisation of the environmental movement.

HARRY ROTHMAN is Professor of Science and Technology Policy in the Bristol Business School, and Co-Director of the Science and Technology Policy Centre (Sci-Tec). He has published over sixty articles and five books in the area of science and technology policy, including *The Biotechnology Challenge* (Cambridge University Press, 1986). He has been a consultant to a number of international organisations including UNESCO and has completed a study of the current state of science and science policy studies in the UK for the Economic and Social Research Council. He is the founding editor of *Technology Analysis and Strategic Management*.

PROFESSOR PETER GLASNER, in addition to co-directing Sci-Tec, is Dean of the Faculty of Economics and Social Science at the University of the West of England, Bristol. He has written a number of books including *Computers, Risks and Hazards* (with D. Travis) published by Routledge. An article on public understanding and the Human Genome Project has been simultaneously published in French and English as part of the 1993 European Week of Science and Culture.

CAMERON ADAMS was the Research Assistant on the project which forms the basis of Chapter Nine. He has also carried out work on the NIMBY syndrome and 'greening business enterprise'. He is now conducting research at the Southampton Institute of Higher Education.

Select bibliography

Public understanding of science is a relatively new field and also one with ill-defined boundaries and many interconnections with other areas of analysis and debate. This highly selective bibliography, which we have restricted to books or book-length works, is intended to give the reader access to some of the most relevant strands of research and debate.

A much-quoted example of the perspective of scientists on the public understanding 'problem' is the London Royal Society's 1985 report (Royal Society, 1985). There are numerous other examples of scientists bemoaning the public 'lack of understanding' and defending the significance of the scientific enterprise – one interesting account within this broad discourse is Wolpert (1992). An opposite view, critical of the scientism embodied in conventional scientific perspectives on public culture, but itself offering rather too-simple a dichotomy between 'the two cultures', is Appleyard (1992). Schwartz (1992) is a more balanced account from a scientist.

Collins and Pinch (1993) attempt to popularise sociology of scientific knowledge (SSK) itself, in a book which uses SSK and several illustrations from scientific practice to argue that public understanding efforts should be focused on what sociologists and others have learnt about science as a process, rather than upon its cognitive contents. Layton, Jenkins, Macgill, and Davies (1993) provide case-studies and analysis close to those of the present collection, attempting to avoid the a priori prescriptions embodied in conventional approaches. The collection by Lewenstein offers a related line of analysis (1992). Myers (1990) provides an interesting basis for thinking about the forms of communication within science, and beyond it to its publics.

In more historical vein, Layton (1973) has critically examined the programmes of public education in science in the nineteenth century, identifying their ideological role in inculcating 'correct' views of much more than just nature itself. This corresponds with more recent work in history of science, for example, Shapin and Schaffer (1985) and Golinski (1992) which pays attention to the ways in which local knowledge, generated in the privacy of the laboratory, was successfully constructed as authoritative 'universal' public knowledge. A more wide-ranging work addressing the gender issue from a broadly similar intellectual perspective is Jordanova's (1989) historical analysis of 'sexual visions'. In highlighting the senses in which social order and scientific culture were co-produced and mutually validated, these studies resonate with the more contemporary work of Latour (1987). Consistent with Latour's idea that the scientific laboratory acts as a key organising point of wider sociotechnical order and power, various authors have begun to examine the role of 'public experiments' and technological 'testing' in society. Some valuable works of this kind are collected in Gooding, Pinch, and Schaffer (1989).

LaFollette (1990) reviews popular images of science during the period of development of the mass media, from 1910 to 1955. Nelkin (1987) discusses the fundamentally uncritical treatment of science by journalists, even when political stories about science are being covered. Friedman, Dunwoody, and Rogers (1986) offer a collection of accounts of the construction of science as news in the print media, whilst Silverstone's (1985) case-study remains one of the few sociological analyses of the construction of science for public consumption on television. Of many interpretive treatments of science and technology in popular culture, Tudor (1989) and Winner (1977) are rewarding. Shinn and Whitley (1985), and Holton and Blanpied (1976) are two collections with case-studies of the underlying politics of science popularisation.

In exploring the relationships between lay and scientific reasoning, an enormous literature exists especially in sociology, anthropology, and cognitive psychology. Holland and Quinn (1987) offer several cognitive psychology analyses of lay 'mental models', whilst Lave (1988) and Rogoff (1990) provide a more anthropological understanding of such cognitive-social processes. In modern political arenas such as risk and environmental controversy, the logic and legitimacy of expert and non-expert frameworks of knowledge have been critically analysed by authors such as, for example, Douglas (1986) and Krimsky and Golding (1992). A topical anthropological study of lay experience and the cultural framing of nuclear technology is Zonabend (1993). In contexts of Third World development and modernisation, similar tensions between lay and scientific reasoning arise. Goonatilake's account in *Aborted Discovery* (1984) argues that modern Western science has overwhelmed the rich scientific tradition of non-European civilisations.

The politics of democratic involvement in science has been persistently analysed by Nelkin in several works (for example, 1992). From the extensive field of public health Burnham (1987) gives a polemical critique of the 'irrationality' of lay thinking, whilst Helman (1991) gives a more complex and sympathetic treatment from the experience of an anthropologically trained doctor. Fischer (1990) gives some examples and analyses attempts at popular participation in science. A somewhat dated, but still useful, review of public participation in technological decision-making is provided by the OECD (1979).

Finally, and at a more theoretical level, the work of Ulrich Beck (1992, 1995) has strong significance for debates around the 'reconstruction' of science within everyday life. Beck's work is also part of a wider sociological debate about the changing nature of modernity (for example Giddens 1991; Bauman 1991, Lash, Szerszynski, and

Wynne 1996) and the role of scientific and 'lay' groups within the new condition of 'late' or 'post' modernity.

REFERENCES

Appleyard, B., 1992, *Understanding the Present; science and the soul of modern man* (London: Pan).

Bauman, Z., 1991, *Modernity and Ambivalence* (Cambridge: Polity Press).

Beck, U., 1992, *Risk Society; towards a new modernity* (London, Newbury Park, New Delhi: Sage).

Beck, U., 1995, *Ecological politics in an Age of Risk* (Cambridge: Polity Press).

Burnham, J., 1987, *How Superstition Won and Science Lost: popularising science and health in the United States* (New Brunswick, N.J.: Rutgers University Press).

Collins, H. and Pinch, T., 1993, *The Golem: what everyone should know about science* (Cambridge and New York: Cambridge University Press).

Douglas, M., 1986, *Risk Acceptance According to the Social Sciences* (New York: Russell Sage Foundation).

Fischer, F., 1990, *Technocracy and the Politics of Expertise* (Newbury Park, London and New Delhi: Sage).

Friedman, S., Dunwoody, S., and Rogers, C. (eds.), 1986, *Scientists and Journalists: reporting science in the news* (New York: Free Press).

Giddens, A., 1991, *Modernity and Self-Identity* (Cambridge: Polity Press, 1991).

Golinski, J., 1992, *Science as Public Culture: chemistry and enlightenment in Britain, 1760–1820* (Cambridge and New York: Cambridge University Press).

Gooding, D., Pinch, T., and Schaffer, S. (eds.), 1989, *The Uses of Experiment: studies in the natural sciences* (Cambridge and New York: Cambridge University Press).

Goonatilake, S. 1984, *Aborted Discovery; science and creativity in the Third World* (London: Zed).

Haraway, D., 1989, *Primate Visions: gender, race and nature in the world of modern science* (New York: Routledge).

Helman, C., 1991, *Body Myths* (London: Chatto and Windus).

Holland, D., and Quinn, N. (eds.), 1987, *Cultural Models in Language and Thought* (Cambridge University Press).

Holton G., and Blanpied W. (eds.), 1976, *Science and its Public; the changing relationship* (Boston: Reidel).

Jordanova, L., 1989, *Sexual Visions; images of gender in science and medicine between the eighteenth and twentieth centuries* (Brighton: Harvester Wheatsheaf).

Krimsky S. and Golding D. (eds.), 1992, *Social Theories of Risk* (Westport, Conn. and London: Praeger).

LaFollette, M., 1990, *Making Science our Own: public images of science 1910–1955* (Chicago and London: University of Chicago Press).

Lash, S., Szerszynski, B., and Wynne, B. (eds.), 1996, *Risk, Environment and Modernity: towards a new ecology* (London: Sage).

Latour, B., 1987, *Science in Action* (Milton Keynes: Open University Press).

Lave, J., 1988, *Cognition in Practice: mind, mathematics and culture in everyday life* (Cambridge and New York: Cambridge University Press).

Layton, D., 1973, *Science for the People. The origins of the school science curriculum in England* (London: Allen and Unwin).

Layton, D., Jenkins, E., Macgill, S., and Davey, A., 1993, *Inarticulate Science?* (Driffield, UK: Studies in Education Ltd).

Lewenstein, B., *When Science Meets the Public* (Washington, DC: AAAS, 1992).

Myers, G., 1990, *Writing Biology* (Madison: University of Wisconsin Press).

Nelkin, D., 1984, *Science as Intellectual Property: who controls research?* (London: Collier MacMillan).

Nelkin, D., 1987, *Selling Science* (New York: Freeman).

Nelkin, D. (ed.), 1992 *Controversy: politics of technical decisions* (Newbury Park, London and New Delhi: Sage).

OECD, 1979, *Technology on Trial* (Paris: OECD).

Rogoff, B., 1990, *Apprenticeship in Thinking: cognitive development in social context* (New York and Oxford: Oxford University Press).

Ross, A., 1992, *Strange Weather: culture, science and technology in an age of limits* (London and New York: Verso).

Royal Society, 1985, *The Public Understanding of Science* (London: Royal Society).

Schwartz, J., 1992, *The Creative Moment: how science made itself alien to modern culture* (London: Jonathan Cape).

Shapin, S. and Schaffer, S., 1985, *Leviathan and the Air Pump: Hobbes, Boyle, and the experimental life* (Princeton University Press).

Shinn, T. and Whitley, R. (eds.), 1985, *Expository Science: forms and functions of popularisation* (Dordrecht: Reidel).

Silverstone, R., 1985, *Framing Science: the making of a BBC documentary* (London: BFI Publishing).

Tudor, A., 1989, *Monsters and Mad Scientists: a cultural history of the horror movie* (London: Basil Blackwell).

Winner, L., 1977, *Autonomous Technology* (Cambridge, Mass.: MIT Press).

Wolpert, L., 1992, *The Unnatural Nature of Science* (London: Faber and Faber).

Zonabend, F., 1993, *The Nuclear Peninsula* (Cambridge University Press).

Index